T0339744

Phenotyping Crop Plants for Physiological and Biochemical Traits

Phenotyping Crop Plants for Physiological and Biochemical Traits

P. Sudhakar
Department of Crop Physiology
S. V. Agricultural College
Acharya N. G. Ranga Agricultural University
Tirupati, A.P., India

P. Latha
Institute of Frontier Technology
Regional Agricultural Research Station
Acharya N. G. Ranga Agricultural University
Tirupati, A.P., India

P.V. Reddy
Regional Agricultural Research Station
Acharya N. G. Ranga Agricultural University
Tirupati, A.P., India

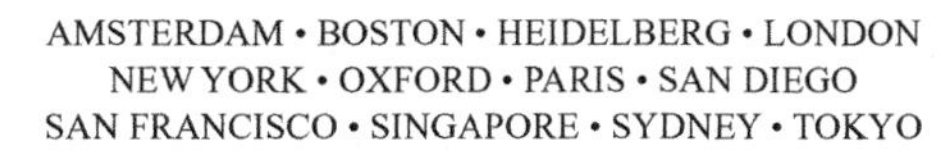

Academic Press is an imprint of Elsevier

Academic Press is an imprint of Elsevier
125 London Wall, London EC2Y 5AS, UK
525 B Street, Suite 1800, San Diego, CA 92101-4495, USA
50 Hampshire Street, 5th Floor, Cambridge, MA 02139, USA
The Boulevard, Langford Lane, Kidlington, Oxford OX5 1GB, UK

Distributed in India, Pakistan, Bangladesh, and Sri Lanka by BS Publications.

British Library Cataloguing-in-Publication Data
A catalogue record for this book is available from the British Library

Library of Congress Cataloging-in-Publication Data
A catalog record for this book is available from the Library of Congress

ISBN: 978-0-12-804073-7

For information on all Academic Press publications
visit our website at https://www.elsevier.com/

Publisher: Nikki Levy
Acquisition Editor: Nancy Maragioglio
Editorial Project Manager: Billie Jean Fernandez
Production Project Manager: Nicky Carter
Designer: Matthew Limbert

Typeset by Thomson Digital

Contents

Message

The growing demand for food and increasing scarcity of fertile land, water, energy, etc., present multiple challenges to crop scientists to meet the demands of future generations while protecting the environment and conserving biological diversity. The productivity of crops greatly depends on the prevailing environmental conditions. Although farming practices are capable of increasing crop yields through control of pests, weeds, and application of fertilizers, the weather cannot be controlled. Occurrence of abiotic stress conditions such as heat, cold, drought, flooding causes huge fluctuations in crop yields. Climatic change scenarios predict that weather extremes are likely to become more prevalent in the future, suggesting that stress proofing our major crops is a research priority.

Crop physiology plays a basic role in agriculture as it involves study of vital phenomena in crop plants. It is the science concerned with processes and functions and their responses toward environmental variables, which enable production potential of crops. Many aspects of practical agriculture can be benefited from more intensive research in crop physiology. Hence, knowledge of crop physiology is essential to all agricultural disciplines that provide inputs to Plants Breeding, Plant Biotechnology, Agronomy, Soil Science, and Crop Protection Sciences.

Novel directions in linking this basic science to crop and systems research are needed to meet the growing demand for food in a sustainable way. Crop performance can be changed by modifying genetic traits of the plant through plant breeding or changing the crop environment through agronomic management practices. To achieve that, understanding crop behavior under environmental variables plays an important role in integrating and evaluating new findings at the gene and plant level. Reliable crop-physiological techniques are essential to phenotype crop plants for improved productivity through conventional and molecular breeding.

The authors of this book have been working on developing various physiological and biochemical traits in different field crops for 20 years and have established state-of-the-art laboratory and field facilities for phenotyping crop plants at Regional Agricultural Research Station, Tirupati. I congratulate the authors for their studious efforts in bringing out their expertise in the form of this book. I hope this book provides an insight into several physiological and biochemical techniques that can benefit scientists, teachers, and students of Agriculture, Plant Biology, and Horticulture.

A. Padma Raju

Foreword

The most serious challenges that societies will face over the next decades are providing food and water, in the face of mounting environmental stresses, warned by the consequences of global climate change. There is an urgent need of developing methods to alleviate the environmental disorders to boost crop productivity especially with existing genotypes, which are unable to meet our requirements.

The Green revolution in cereals promoted optimism about the capacity of crop improvement in increasing yield and it drove plant physiologists to understand the physiological basis of yield and its improvement. Although research in crop physiology encompasses all growth phenomena of crop plants, only traits that have a likely economic impact and show significant genetic variation can be considered in the context of crop improvement.

The first step to be taken in this direction is to use appropriate screening techniques to select germplasm adapted to various abiotic stress conditions. The improvement of abiotic stress tolerance relies on manipulation of traits that limit yield in each crop and their accurate phenotyping under the prevailing field conditions in the target population of environments.

Agricultural scientists and students often face impediments in selecting right phenotyping method in various crop experiments. There is a dire need to bring reliable protocols of physiological and biochemical traits which directly or indirectly influences final yield in a book form. I am well aware that authors of this book Dr P. Sudhakar, Dr P. Latha, and Dr P.V. Reddy have played key role in developing drought-tolerant peanut varieties in this University by applying various physiological traits standardized in their laboratory. I congratulate the authors for bringing out their expertise in the form of this book "Phenotyping crop plants for physiological and biochemical traits."

This publication not only is the detailed explanation of methodology of phenotyping but also links the physiology to a possible ideotype for its selection. Hence, this book is highly useful to agricultural scientists, molecular biologists, and students to select desirable ideotype for their target environment.

K. Raja Reddy

Preface

This book elaborates methods that can contribute to phenotyping of crop plants for various physiological and biochemical traits. It contains field-based assessment of these traits, as well as laboratory-based analysis of tissue constituents in samples obtained from field-grown plants. Most of the phenotyping methods given in this book are reliable, as they were validated in our research programmes.

We extend thanks to all the colleagues for their support in validating the phenotyping methods in several agricultural crops. We express deep sense of reverence and indebtedness for all the team members of this crop physiology department since 1996, viz., Narsimha Reddy, D. Sujatha, Dr M. Babitha, Dr Y. Sreenivasulu, Dr K.V. Saritha, B. Swarna, M. Balakrishna, T.M. Hemalatha, V. Raja Srilatha, C. Rajia Begum, and K. Lakshmana Reddy. We appreciate K. Sujatha, Senior Research Fellow of this department, for her involvement in validating phenotyping methods as well as in preparation of this book.

We express gratitude for Dr T. Giridhara Krishna, Associate Director of Research, Regional Agricultural Research station, Tirupati and Dr K. Veerajaneyulu, University Librarian for their constant support in accomplishing this book. We are grateful to Acharya N G Ranga Agricultural University for facilitating the research needs and support in bringing out this book.

We extend special thanks to our collaborate scientists Dr S.N. Nigam, ICRISAT, Dr M. Udaya Kumar, UAS, Bangalore, Dr R.C. Nageswara Rao, ACIAR, Australia, and Dr R.P. Vasanthi, RARS, Tirupati for their support over all these years.

Finally, we hope this book provides insightful information about various reliable phenotyping methods adopted in laboratory, greenhouse, and field-oriented crop research for students and researchers of Agriculture, Horticulture, Molecular biology, Botany, and Allied sciences.

- Authors

Abbreviations

cm	Centimeter
mm	Millimeter
°C	Degree centigrade
Δ	Difference
α	Alpha
β	Beta
γ	Gamma
D.H_2O	Distilled water
D.D H_2O	Double distilled water
fr.wt	Fresh weight
g	Gram
GLC	Gas liquid chromatography
h	Hour
HPLC	High-performance liquid chromatography
kg	Kilogram
L	Liter
μCi	Micro curie
μg	Microgram
μL	Microliter
μmole	Micromole
mg	Milligram
min	Minute
mL	Milliliter
mmole	Milli mole
M	Molar
Mol	Mole
N	Normality
nm	Nanometer
OD	Optical density
rpm	Resolutions per minute
s	Second
TLC	Thin layer chromatography
V/V	Volume/volume
W/V	Weight/volume
Y	Year

Introduction

Agricultural crops are exposed to the ravages of abiotic stresses in various ways and to different extents. Unfortunately, global climate change is likely to increase the occurrence and severity of these stress episodes created by rising temperatures and water scarcity. Therefore, food security in the 21st century will rely increasingly on the release of cultivars with improved resistance to drought conditions and with high-yield stability (Swaminathan, 2005; Borlaug, 2007).

We are using landraces as genetic sources for abiotic stress resistance. These are the simple products of farmers who repeatedly selected seed that survived historical drought for years in their fields. No science was involved, only a very long time and a determination to provide for their own livelihood. These landraces attend to the fact that abiotic stress resistance has been here for a very long time. We are now only trying to improve it more effectively.

Improving the genetic potential of crops depends on introducing the right adaptive traits into broadly adapted, high-yielding agronomic backgrounds. The emerging concept of newly released cultivars should be genetically tailored to improve their ability to withstand drought and other environmental constraints while optimizing the use of water and nutrients. A major recognized obstacle for more effective translation of the results produced by stress-related studies into improved cultivars is the difficulty in properly phenotyping relevant genetic materials to identify the genetic factors or quantitative trait loci that govern yield and related traits across different environmental variables.

The Green Revolution in cereals promoted optimism about the capacity of plant breeding to continue increasing yield and it drove plant physiologists to understand the physiological basis of yield and its improvement. The physiological basis of the Green Revolution in the cereals was identified very early as an increase in harvest index from around 20–30% to about 40–50%, depending on the crop and the case. The yield components involved in this increase were also identified, with grain number per inflorescence as the primary one. Crop physiology then led breeders to understand that yield formation in cereals is derived from an intricate balance between yield components' development, source to sink communication, crop assimilation, and assimilate transport linked to crop phenology and plant architecture (Tuberosa and Salvi, 2004).

Taking full advantage of germplasm resources and the opportunities offered by genomics approaches to improve crop productivity will require a better understanding of the physiology and genetic basis of yield adaptive traits. Although research in plant physiology encompasses all growth phenomena of healthy plants, only traits that have a likely economic impact and which show significant genetic variation can be considered for improvement in the context of plant breeding. Many such traits are expressed at the whole plant or organ level.

Plants exhibit a variety of responses to abiotic stresses, in other words, drought, temperature, salt, floods, oxidative stress which are depicted by symptomatic and quantitative changes in growth and morphology. The ability of the plant to cope with or adjust to the stress varies across and within species as well as at different developmental stages. Although stress affects plant growth at all developmental stages, in particular anthesis and grain filling are generally more susceptible. Pollen viability, patterns of assimilates partitioning, and growth and development of seed/grain are highly adversely affected. Other notable stress effects include structural changes in tissues and cell organelles, disorganization of cell membranes, disturbance of leaf water relations, and impedance of photosynthesis via effects on photochemical and biochemical reactions and photosynthetic membranes. Lipid peroxidation via the production of ROS and changes in antioxidant enzymes and altered pattern of synthesis of primary and secondary metabolites are also of considerable importance.

Phenological traits, that is, pheno-phases of the growth and development, have the greatest impact on the adaptation of plants to the existing environment all with the aim of achieving a maximum productivity (Passioura, 1996). The extent by which one mechanism affects the plant over the others depends upon many factors including species, genotype, plant stage, composition, and intensity of stress.

Phenotype (from Greek *phainein*, to show) is the product of all of the possible interactions between two sources of variation, the genotype, that is, the genetic blueprint of a cultivar, and the environment, that is, the collection of biotic, abiotic, and crop management conditions over which a given cultivar completes its life cycle. Therefore, even discrete observations of a given phenotype can integrate many genotype and environmental connections over time. Genotype-by-environment interactions can play a significant role in the phenotypes collected in the field or greenhouse.

Phenotyping involves measurement of observable attributes that reflect the biological functioning of gene variants (alleles) as affected by the environment. To date, most phenotyping of secondary traits (ie, those traits in addition to yield, the primary trait) has involved field assessments of easily scored morphological attributes such as plant height, leaf number, flowering date, and leaf senescence. However, phenotyping plants for abiotic stress tolerance involves metabolic and regulatory functions, for which measurements of targeted processes are likely to provide valuable information on the underlying biology and suggest approaches by which it could be modified.

Good phenotyping is a critical issue for any kind of experimental activity, but the challenges faced by those investigating the abiotic effects on crops are particularly daunting due to difficulties in standardizing, controlling, and monitoring the environmental conditions under which plants are grown and the data are collected, especially in the field. Phenotypic traits need to be adopted also depending on whether the experiments are carried out in the field or in the controlled environment of a growth chamber or greenhouse. Phenotyping means not only the collection of accurate data to minimize the experimental error introduced by uncontrolled environmental and experimental variability, but also the collection of data that are relevant and meaningful from a biological and agronomic standpoint, under the conditions prevailing in farmers' fields.

Collecting accurate phenotypic data has always been a major challenge for improvement of quantitative traits. Success of this task is intimately connected with heritability of the trait, namely portion of phenotypic variability accounted for by additive genetic effects that can be inherited through sexually propagated generations (Falconer, 1981). Trait heritability varies according to the genetic makeup of the materials under investigation, the conditions under which the materials are investigated, the accuracy and precision of the phenotypic data. Despite this, careful evaluation and appropriate management of the experimental factors that lower the heritability of traits, coupled with a wise choice of the genetic material, can provide effective ways to increase heritability and hence the response to phenotypic selection.

Moreover, excellent methods have been developed for assay of such traits and they have been used in controlled studies to determine the mechanistic basis of stress response. Notwithstanding their positive aspects, these methods often require highly controlled laboratory environments and are too time consuming and expensive or technically demanding to be used in large-scale phenotyping.

The challenge, then, is to identify those attributes that provide the most meaningful phenotypic information, to design sampling methods suitable for use in the field, and to design analytical methods that can efficiently be scaled up to the number of samples required for phenotyping of crops in field experiments. Selection for one trait can reduce a chance for a successful selection for some other trait, due to a competitive relationship toward the same source of nutrients. However, the combination of traits that in various ways contribute to the improvement of yields can result in a maximum gain of each trait individually.

Although earlier studies reported several physiological and molecular traits with the relevance field applicability, many of them are not simple, reliable, and researcher friendly due to complicated protocols and high genotype and environment interaction. This book will discuss various methods that can contribute to phenotyping of crop plants for various physiological and biochemical traits. They involve analyzing methods for field-based assessment of these traits, as well as laboratory-based analyses of tissue constituents in samples obtained from field-grown plants. Researchers or students working in this direction will have several options to select the reliable methodology according to the objective and experimenting conditions.

SECTION I

CHAPTER

Various methods of conducting crop experiments

1

Effective phenotyping should require a set of core setups in which plants are cultivated either under laboratory conditions or in experimental fields. Such experiments enable researchers to determine the phenotypic responses of plants to defined experimental treatments and evaluate the performance of different genotypes or species in a given environment. To enable generalizations across experiments, it is necessary that results are not only replicable, but also reproducible. Replication of results is achieved when the same researcher finds the same results while repeating an experiment in time. In plant biology, achieving a high degree of reliability and reproducibility is a challenge. This chapter provides information on different methods of conducting experiments for crop and data to be recorded on various abiotic environment parameters apart from regular plant biometric data.

1.1 FIELD EXPERIMENTS

Field experiments with rainout shelter facility are shown in Fig. 1.1. These are typically undertaken under conditions where some, but not all variables, can be controlled. These sometimes represent a particular stress (eg, drought, nutrient, or temperature), or under favorable conditions where the aim is to understand physiological and agronomic factors contributing to yield potential. Similarly, assessment of genotypes under a controlled stress requires an understanding and reporting of factors contributing to their differential performance in response to stress. If some of the observed differences in yield relate to differences in plant height, flowering, or greater leaf area, then the cataloguing of such variation must be undertaken.

Measuring and reporting of this variation can be varied among the researchers. This makes interpretation across multiple experiments difficult as one researcher may view and undertake sampling differently from another. It is critical that there is consistency in how measurements are undertaken and reported. Hence, standardizing procedures and phenotyping among individuals will provide data that are robust, reliable, and repeatable. This will lead to more cost-efficient research wherein high-quality data can be produced and reused.

1. *Selection of site*: For critical planning and interpreting field response data, good knowledge of the site and expected seasonal conditions based on prior knowledge of long-term weather trends are essential. Information such as soil

Phenotyping Crop Plants for Physiological and Biochemical Traits. http://dx.doi.org/10.1016/B978-0-12-804073-7.00001-6

FIGURE 1.1 Field Experimentation With Rainout Shelter Facility.

conditions, viz., soil type, soil texture, soil moisture, soil nutritional status, soil born pest, and diseases, should be analyzed for the experiment area. Identification of uniform blocks with perfect leveling to reduce residual (error) variation in large size field experiments is essential. Long-term seasonal rainfall and temperatures are to be collected and should be used in planning for the need for sowing date, irrigation, and imposing abiotic stresses.

2. *Plot type and size*: Phenotyping of complex physiological traits and particularly hose associated with canopy development, biomass, and yield is challenging when experiments comprise diverse genotypes. This is especially so when confounded with variation in traits such as height and maturity that are known to affect yield.
3. *Implications in row and plot experiments*
   a. Row plantings
   Limited seed and resources may encourage field assessment in single, spaced rows or smaller, unbordered plots. Competition for water, light, and nutrients required for canopy growth is variable as adjacent rows are genetically different and competition is greatest particularly under stress conditions. Response to changes in resource availability varies among diverse genotypes, alters genotype ranking, and thus reduces heritability. In turn, the relevance of such growing conditions to commercial field-grown crops is unclear.
   b. Plot experiments
   The planting of multirow plots and the simple exclusion of plot borders at harvest increases experimental precision and confidence in genotype

response. Well-planned field studies and particularly those focusing on the dynamics of yield formation (canopy-related characteristics) must consider the use of multiple-row plots and with border rows to minimize the effects of inter-plot competition. Plots should contain two outer rows ("edge" or "border") and multiple inner rows to minimize inter-plot competition effect, for example, edge effects due to shading, nutrients, water availability, or compaction.

c. Phenotyping in the field
Assuming that both the type and the number of treatments (genotypes, irrigation volumes, etc.) to be evaluated are adequate for the specific objectives of each experiment, the following general factors should be evaluated carefully to ensure the collection of meaningful phenotypic data in field experiments conducted under water-limited conditions:

- experimental design
- heterogeneity of experimental conditions between and within experimental units
- size of the experimental unit and number of replicates
- number of sampled plants within each experimental unit
- genotype-by-environment-by-management interaction.

4. *Weather measurements*
The weather has a huge impact on the crop growth and development, and the stress that the plants will experience. Recording accurately the main weather variable is thus crucial in success of any field experiment.

a. Stable weather station
Generally, the daily weather data, viz., solar radiation, rain, maximum and minimum temperatures, wind speed, air humidity, pan evaporation, are collected from research stations where experiment is conducted or nearby organizations that have stable weather station. The demerits of such data are

- They can be far from the field trial, whereas environmental factors such as rain can vary within short distances.
- They only deliver daily measurements that are not always accurate to evaluate stress events.

b. Portable weather station
A better alternative is to install a portable weather station in the field trial, to record climatic data more frequently (eg, measurements every minute). Typically, these weather stations have a solar radiation sensor, a tipped-bucket rainfall gauge, and an air temperature and relative humidity probe mounted in a Stevenson screen. In addition, many other sensors can also be included, such as:

- Thermistors to measure soil temperature
- Thermocouples to measure soil, leaf temperatures
- Infrared sensors to measure canopy temperature continuously
- Solarimeter tubes to measure light interception
- An anemometer and a wind vane to measure wind speed and direction.

5. *Merits of field experiments*
    a. Field conditions are relatively close to the natural environment that crops experience in the field.
    b. It provides an opportunity to compare plants under conditions in which spatial heterogeneity is relatively small.
6. *Demerits of field experiments*
    a. Uncontrolled variations in light, temperature, and water supply.
    b. Various environmental conditions may change in concert, that is, a period of high irradiance may come with high temperatures and low precipitations.

1.2 EXPERIMENTS UNDER GREEN HOUSES

Glasshouses and polyhouses are good alternative and provide more buffered conditions for growing plants. They offer better control of water supply and protection against too low temperatures. Additional lighting in the glasshouse may ensure a minimal daily irradiance and a fixed photoperiod, whereas shade screens can protect against high light intensities in summer (Max et al., 2012).

1.2.1 DEMERITS

1. In practical terms, plants grown in glasshouses will usually experience higher-than-outdoor air temperatures during nights and winters and lower irradiance because of shading.
2. Most glasshouses or polyhouses without humidity control have limited possibilities of reducing temperatures during periods of strong solar irradiance in summer.
3. In many greenhouses where there is no artificial lighting, significant spatial heterogeneities in irradiance due to shading by the greenhouse structure itself are observed.

1.3 EXPERIMENTS IN GROWTH CHAMBERS

Climate-controlled growth chambers (Fig. 1.2) are expensive in terms of investments as well as running costs. They offer the most sophisticated possibilities for environmental control and thereby good reliability of experiments.

1.3.1 DEMERITS

1. Conditions in growth chambers are generally the furthest away from those in the field, not only because environmental values are often programmed within a relatively small diurnal range, but also with regard to the absolute values of, for example, light and temperature, at which they operate (Garnier and Freijsen, 1994).

FIGURE 1.2 Climate-Controlled Growth Chamber.

2. Although growth chambers enable a strong temporal control over conditions, spatial variability is often larger than anticipated and higher than those measured in experimental fields. For example, light intensity may vary from place to place in the growth chamber (Granier et al., 2006) and can be especially lower close to the walls.
3. Gradients in air velocity may go unnoticed in growth chambers, although they can affect evaporative demand. Variation in air circulation may be especially large when plant density is high or plants are placed in trays, which may block air circulation around the plants. Both too high and too low wind speeds are undesirable.
4. A factor that may strongly vary in a temporal manner is the local atmospheric CO_2 concentration; generally, CO_2 levels in a building are higher than outside.
5. Under greenhouses as well as growth chambers crops are experimented through either hydroponics or pot culture method of growing crops.

1.4 HYDROPONICS

Roots provide nearly all the water and nutrients that a plant requires. If the aim is to design an experiment in which these two factors have the least limiting effect on growth, then hydroponics or aeroponics is the preferred choice (Gorbe and Calatayud, 2010). Hydroponics systems can be either based on roots suspended in a water

FIGURE 1.3 Hydroponics Experiment.

solution or in some solid medium such as sand, rockwool, or another relatively inert medium, which is continuously replenished with nutrient solution (Cooper, 1979). Frequently used nutrient solutions were described by Hoagland and Snijder (1933) and Hewitt (1966), although the truly optimal composition is species specific. Preparation of macro, micronutrients (Appendix VI) and Hoagland solution (Appendix VII) were given in appendices as ready recoknoire. Hydroponics experiment is shown in Fig. 1.3.

1.4.1 PRECAUTIONS

1. Water-based systems have the advantage that they allow easy experimental access to the roots for physiological or biomass measurements. However, care has to be taken while roots are transferred from one solution to another, as breakage of roots may easily occur.
2. There is also a need to take into account the composition of tap water when setting for the final composition. Because of the much higher mixing rate in soilless systems and the direct access of plant roots to the nutrients, the concentrations of nutrients that are needed to sustain supply are 5–10 times lower than those required for plants growing on sand where there is an absence of continuous flow through.

3. Ensure that the concentration of macro and especially micronutrients in a hydroponics system is not too high, as this will negatively affect plant growth or may even cause leaf senescence (Munns and James, 2003). On the other hand, nutrient concentrations should not become too low either, as plants will otherwise deplete the available minerals. Hence, regular replacement of nutrient solution is necessary.
4. Bigger plants usually need more nutrients and so the rate of replenishment must increase with plant size, unless the nutrient concentration itself is continuously monitored and adjusted.
5. Good mixing of aerated nutrient solution is vital to avoid depletion zones around the roots and anaerobic patches, but should not be too vigorous to avoid strong mechanical strains. In addition, specific uptake mechanisms such as the release of chelating agents to increase iron availability (Romheld, 1991) or the release of organic acids by the root may be affected.
6. The pH of the hydroponic solution may increase or decrease, depending on whether nitrate or ammonium is present in the solution and the specific preference of a given species. For most plant species a pH of 6 seems to be optimal, although specific species may deviate significantly. Monitoring and adjusting the pH of the solution at a regular basis is highly recommended, keeping in mind that pH changes are stronger in small volumes of nutrient solution and for roots with faster nitrogen uptake rates.
7. It should also be checked that there is no accumulation of salts at the root: shoot junction over time, as this can damage the seedlings of some plant species.

1.5 POT CULTURE

An alternative to hydroponics is to grow plants in pots filled with an inert solid medium (eg, sand, perlite) or soil and to water them regularly or on demand. Use of pots with a solid substrate may at least mimic the higher mechanical impedance to root growth that plants experience in soils and allows for a higher homogeneity and control of the nutrient and water conditions than in soil. Pot culture (Fig. 1.4) allows more freedom in the choice of the location of the experiment and ensures

FIGURE 1.4 Pot Culture Experiment.

easy handling and manipulation of the shoots of individual plants. Most overlooked factors in pot culture are pot size and the fact that nutrients and water supply strongly interact with plant size.

1. *Pot size*: The size of the rooting volume also requires careful attention. The smaller the pot, the more plants fit into a growth chamber or glasshouse, an advantage for nearly all laboratories where demand for space is high. At the same time, in most experiments smaller pots will also imply a lower availability of below-ground resources and if pots are closely spaced, also a comparatively lower amount of irradiance available for each plant. Moreover, the smaller the pot the stronger roots become pot-bound, leading to undesirable secondary effects. In experiments in which rooting volume varies, there is almost invariably a strong positive correlation between plant growth and pot volume reported. Conditions obviously differ from experiment to experiment, but as a rule of thumb, pot size is certainly small if the total plant dry mass per unit rooting volume exceeds 2 g/L (Poorter et al., 2012).
2. *Precautions*:
   a. Demands for water and nutrients increase strongly with the size of the plants, so the water and nutrient availability that are amply sufficient for small plants at an early phase may become limiting at later developmental stages.
   b. Nutrient availability of commercially provided soil will vary among suppliers and even over time from soil batch to soil batch. Mixing of slow-release fertilizer with the soil or regular addition of nutrient solution may mitigate this problem to some extent.
   c. Root damage may occur if pots are black and warm up under direct solar radiation. Moreover, soil temperature per se and even gradients in soil temperature within single pots can affect plant growth and allocation (Fullner et al., 2012).

Phenotyping experiments with plants require careful planning. The most controlled growth environment is not necessarily always the best one. Growing crop plants for experimental purposes remains an art, requiring in-depth knowledge of physiological responses to the environment together with proper gauging of environmental parameters. Hence, it is advocated to adopt a practical checklist (Table 1.1) to document and report an asset of information concerning the abiotic environment, plants experienced during experiments. Similarly, advantages and disadvantages of field versus controlled environments in relation to some physiological traits are given in Table 1.2.

Table 1.1 Checklist With the Recommended Basic and Additional Data to Be Collected in All Methods of Experimentation

Sr. No.	Basic Data	Additional Data
1. Light intensity (PAR)	• Average daily integrated PPFD measured at plant or canopy level (mol m^{-2} day^{-1}) • Average length of the light period (h)	• For GC: light intensity (μ mol m^{-2} s^{-1}) • Range in peak light intensity (μ mol m^{-2} s^{-1}) • For GH: fraction of outside light intercepted by growth facility components and surrounding structures
2. Light quality	• For GC and GH: type of lamps used	• R/FR ratio (mol mol^{-1}) • Daily UV-B radiation (W m^{-2}) • Total daily irradiance (W m^{-2})
3. CO_2	• For GC and GH: controlled/ uncontrolled	• Average [CO_2] during the light and dark period (μ mol mol^{-1})
4. Rooting medium	• Water-based hydroponics/ solid-based hydroponics including substrate used/soil type • Container volume (L) • Number of plants per container • For hydroponics and soil: pH • Frequency and volume of replenishment or addition	• Container height • For soil: soil penetration strength (Pam^{-2}); water retention capacity (g g^{-1} dry weight); organic matter content (%); porosity (%) • Rooting medium temperature
5. Nutrients	• For hydroponics: composition • For soil: total extractable N before fertilizer added • For soil: type and amount of fertilizer added per container or m^2	• For soil: concentration of P and other nutrients before start of the experiment • For soil: total extractable N at the end of the experiment
6. Air humidity	• Average VPD air during the light period (kPa) or average humidity during the light period (%)	• Average VPD air during the night (kPa) or average humidity during the night (%)
7. Water supply	• For pots: volume (L) and frequency of water added per container or m^2 • Average day and night temperature (°C)	• For soil: range in water potential (MPa) • For soil: irrigation from top/ bottom/drip irrigation • Changes over the course of the experiment
8. Salinity	• Composition of nutrient solutions used for irrigation	• For hydroponics: composition of the salts (mol L^{-1}) • For soils and hydroponics: electrical conductivity (dS m^{-1})

GC, growth chamber; GH, glass house.
Adapted from Poorter et al. (2012).

Table 1.2 Advantages and Disadvantages of Field Versus Controlled Environments in Relation to Some Physiological Traits

	Field		Controlled Facilities	
Traits to Study	**Advantages**	**Disadvantages**	**Advantages**	**Disadvantages**
Treatments	Realistic	Less uniform	Control of the intensity, uniformity, timing, and repeatability of treatments.	Unrealistic
		Dependence on environmental/seasonal factors	Out-of-season experiments are possible	
		Unpredicted interactions	Interactions between factors can be controlled	Variation in the glasshouse environment and handling of materials
			Particular variables (radiation, ozone, etc.) can be manipulated and monitored	
Responses to drought	Realistic drying cycles	Cooccurrence of additional stresses (heat, low temperature)	Control of environmental factors	Unrealistic (rapid) drying cycles
	Realistic interactions with environmental factors	Less control over treatments	Control of water applied	Confounded by plant growth rate and differences in water status
	Realistic soil profile for root development	Confounding factors (toxicities, salinity)		Pot experiment limitations on root growth
Osmotic adjustment		Confounded by root depth and differences in soil water potential	Control of root depth	Unrealistic (rapid) drying/ rehydration cycles
			Equal soil water potential by growing all genotypes in the same pot	

Transpiration efficiency		Water fluxes cannot be controlled	Precise control of water fluxes	
Canopy temperature	Integrative measurement, scoring the entire canopy of many plants Related to the capacity of the plants to extract water from deeper soil profiles	Measurements must be taken when the sky is clear and there is little or no wind	Control of external factors	Only single plant/small groups of plants can be screened Not related to the capacity to extract water from deeper soil profiles -unless special pots are used
Root growth studies	Realistic soil profile	Heterogeneity	Complete root systems are collected	Pot size, temperature, salinity, and hypoxia limiting root growth
(Biomass, length, growth rate, etc.)		High sampling variance	Uniform sampling	
Adaptation to harsh soil	Realistic	Soil properties difficult to manipulate	Soil properties can be manipulated	Unrealistic
Phenotyping	Realistic	Risk of pollen flow	Low risk of pollen flow	Pot experiment limitations
Transgenic plants		Strict regulations and protocols	Less/easier regulations	

Adapted from Reynolds et al. (2012).

SECTION

II

CHAPTER

Seed physiological and biochemical traits

2

Seed is the basic input in agriculture. It differs from other inputs in terms of having life. Hence, scientific methods are involved in producing and storing it. Maintenance of seed quality is mandatory in selling seed lots. Seed lots are evaluated on the basis of their germination capabilities and vigor. Both germination and vigor of a plant depend on the environment to which plant is exposed, especially from grain filling stage. Genotypic variability in vigor and initial seedling establishment was noticed among crop genotypes. Hence, several physiological and biochemical methods of evaluating crop seed for viability and vigor are described in this chapter.

2.1 DESTRUCTIVE METHODS

2.1.1 SEED VIABILITY

Seed viability is the ability of seed to germinate and produce "normal" seedlings. In another sense, viability denotes the degree to which a seed is alive, metabolically active, and possesses enzymes capable of catalyzing metabolic reactions needed for germination and seedling growth.

2.1.1.1 Seed viability tests

1. *Tetrazolium test*: This test is often referred to as quick test since it can be completed within hours. The test is usually based on measuring the activity of dehydrogenase enzyme in the tissues of embryo. It is conducted by using 2, 3, 5-triphenyl tetrazolium chloride (TTC) solution.

   Principle

   Any living tissue must respire. In the process of respiration the enzyme dehydrogenase will be in a highly reduced state. When the seed is treated with the colorless tetrazolium solution, the living tissue of the seed by virtue of respiration and having the dehydrogenase enzyme in a highly reduced state gives off hydrogen ions. These hydrogen ions reduce the colorless tetrazolium solution into red colored formazan. Thus, the tetrazolium test distinguishes between viable and dead tissues of the embryo on the basis of their relative rate of respiration in hydrated state.

$$\begin{array}{lcl} \text{2, 3, 5 - Triphenyl tetrazolium chloride} & \rightarrow & \text{Triphenyl formazan} + \text{HCl} \\ \text{(colorless)} & & \text{(red color)} \\ \text{oxidized state} & & \text{reduced state} \end{array}$$

Phenotyping Crop Plants for Physiological and Biochemical Traits. http://dx.doi.org/10.1016/B978-0-12-804073-7.00002-8

Procedure

a. Seeds are first soaked in water under ambient conditions to allow complete hydration of all tissues (soak the seed in water for specific time period, viz., for cereals: 3–4 h; for legumes: overnight). Sometimes, a respiration stimulant such as H_2O_2 is added during imbibitions to hasten respiration.
b. For many species the tetrazolium salt can be added to the intact seed. For other seeds cutting or puncturing to permit the accesses of tetrazolium to the parts of the seed is followed.
c. After hydration the seeds are placed in tetrazolium solution (0.1–1.0%) at 25–40°C in dark for 1–8 h. If the seeds are placed for long time, they become over stained and difficult to interpret.

Evaluation

a. Although the tissues of living seeds stain red, estimation of viability requires skill and experience. Sound embryo tissues absorb tetrazolium slowly and tend to develop a lighter color than embryos that are bruised, aged, and frozen or disturbed other ways.
b. The areas of cell division and embryos are most critical during germination. If they are unstained or abnormally stained, the seed germination is weakened.
c. Observe the seeds and count the number of stained and unstained seeds.

2. *Sulfuric acid test*: This is usually a nonenzymatic test. The principle involved in this test is to distinguish the differential coloration of live versus dead tissue when exposed to sulfuric acid. The living portion of the cut surface of the seed develops deep rose color within 5–10 min. Though this test takes less time, one must be careful in handling concentrated sulfuric acid.

2.1.2 SEED VIGOR TESTS

1. *Lab evaluation study:*
   a. *Seed germination test*: Place known number of seeds on a paper towel at an equidistant spacing and cover them by another moistened paper towel and roll the towels in such a way as to keep the seeds intact in their positions between the towels. Then, the rolls are fastened with rubber bands and incubate at a room temperature of 25°C and 100% relative humidity in a seed germinator. After 7 days, open the rolls and examine the seedlings. The counted normal seedlings are expressed as percentage of germination.
   b. *Seedling vigor index (SVI)*: The total length of seedlings is also measured from above germination test after 7 days and the data are used in calculation of seedling vigor index (SVI). It is calculated by using the formula described by Abdul-Baki and Anderson (1973).

$$\text{Seedling vigor index} = \text{Total length (root + shoot) of seedling} \times \text{germination percentage}$$

To generate data on distribution of germination over a time period, 25 seeds from each treatment are arranged in petriplates containing moist filter paper and record daily germination counts. Resumption of radical is considered for germination counts. The data are used to calculate the following parameters (Begum et al., 1987).

a. Mean days of germination:
Mean days of germination = $(t \times n) \times n$
where,
t is the time in days starting from day 0
n is the number of seeds completing germination on day "t."

b. Coefficient of velocity of germination:

$$\text{Coefficient of velocity of germination (CV)} = \frac{N_1 + N_2 + \ldots N_x}{N_1T_1 + N_2T_2 + \ldots + N_xT_x} \times 100$$

where N is the number of seeds germinated at days or time T.

c. Peak value:

$$\text{Peak value} = \frac{\text{Final germination percentage}}{\text{No. of days required to reach that value (peak days)}}$$

d. Mean daily germination:

$$\text{Mean daily germination} = \frac{\text{Final germination percentage}}{\text{Total number of days in a test}}$$

e. Germination value:

$$\text{Germination value (GV)} = \text{Peak value} \times \text{Mean daily germination}$$

2. *Field evaluation study*
Emergence counts: Record daily emergence counts from third day after sowing till the emergence is completed and calculate the following parameters.

a. *Emergence percentage*
Emergence is the numerical ratio of normally emerged seeds to the total number of seeds sown which is expressed in percentage.

b. *Coefficient of rate of germination (CRG) (Heydecker, 1974)*

$$\text{Coefficient of rate of germination} = (n / Dn) \times 100$$

where "n" is the number of seeds emerged on day D and "Dn" is the number of days from the day of sowing. The higher the value the shorter is the time for germination.

c. *Emergence index*
The total of rate of emergence is calculated as emergence index (Baskin, 1969).

$$\text{Emergence index} = \frac{n_1}{dn_1} + \frac{n_2}{dn_2} + \ldots + \frac{n_x}{dn_x}$$

where
n_1 is the number of seeds emerged on the day (first day of appearance of seedlings)
dn_1 is the number of days from the day of sowing
n_x is the number of seeds emerged at the final day
dn_x is the number of days to final count

3. *Biochemical tests related to seed viability and vigor*
   a. *Membrane permeability*: One of the early events associated with seed deterioration in seed stored in ambient condition is loss of membrane integrity. There will be solute loss from seeds when they are placed in water and this would be due to the impaired property of semipermeability. The permeability of cell membrane of seed stored under ambient conditions increases in terms of leaching electrolytes, sugars, and amino acids.
      Membrane permeability of seeds was assessed by analyzing the seed leachate for various kinds of solutes like (1) electrolyte, (2) water-soluble sugars, and (3) water-soluble amino acids.
      - *Electrical conductivity (EC) of seed leachates*
        A quantity of 5 g seeds is taken in 25 mL of distilled water and is allowed to leak. Measure electrical conductivity of the leachates using digital EC meter for every 1 h till constant readings was obtained (Dadlani and Agrawal, 1983a). Values are expressed in Simens per meter (S m^{-1})
      - *Water-soluble sugars*
        After taking the final reading of EC, take 0.5 mL of seed leachate in a test tube and to it add 1 mL of 15% phenol solution followed by 5 mL of concentrated sulfuric acid. Mix the contents thoroughly and cool at room temperature. Read the intensity of color in a spectrophotometer at 490 nm. Prepare a standard curve by using 10–100 mg of glucose per milliliter of solution. The amount of WSS present in the leachate can be derived from the standard curve (Dadlani and Agrawal, 1983a).
      - *Water-soluble amino acids (albumins)*
        For this purpose, take 1 mL of seed leachate in a test tube and add 1 mL of distilled water and 1 mL ninhydrin reagent. Heat the mixture in a water bath at boiling point for a period of 20 min. Later the contents of each tube are diluted with 50% ethanol and are cooled under running water. Read the absorbance of the contents in spectrophotometer at 570 nm. Amount of protein fraction (albumins) in sample is measured by using standard curve prepared with glycine (Plummer, 1988).
   b. *Dehydrogenase activity (Kittock and Law, 1968)*
      The activity of dehydrogenase enzyme is related to the status of vigor in seeds. Seeds stored under ambient conditions lose viability and vigor due to loss of ability to synthesize certain enzymes, among them dehydrogenase is significant.

Soak known number of seeds in distilled water for 16 h in replications. Then, the seeds are cut longitudinally through the embryo into two halves. Later 0.25% aqueous solution of 2,3,5-triphenyl tetrazolium chloride is added to the beaker, containing the cut halves of the seeds. The beakers are kept at 40°C for 4 h. The seeds after turning into red color are washed well under tap water and then 6 mL of 2 methoxy ethanol is added to each beaker containing half seeds. After 10 h when all the stain was extracted from the seeds, read the optical density of colored solution at 470 nm.

c. *Lipid peroxidation*

To assess physiological deterioration of seed, malondialdehyde (MDA) content is determined as a measure of peroxidation (Dadlani and Agrawal, 1987). Seeds of similar moisture content are kept for germination over moist filter paper at 20–25°C for about 24 h. The embryonic axes are excised from the germinating seeds. Ten embryonic axes are homogenized with 7 mL of water. Later 7 mL of TBA-TCA reagent (Thiobarbituric acid – Trichloro acetic acid) is added to it and incubated at 95°C in an oven for 30 min in capped reaction tubes. Afterward the extract is cooled in an ice bath and centrifuged at 5000 g for 10 min. The supernatant is carefully collected and the color intensity is read at 535 nm (a) and 600 nm (b). The difference in the value gives the range of color due to the activity of MDA.

2.2 NONDESTRUCTIVE METHODS

Current technology used to predict seed viability is destructive and results are subjective. Leakage of electrolytes and other solutes during the early phases of germination can be measured from intact seeds and provides rapid, objective data (AOSA, 1983). However, seeds are generally soaked (submerged) in water, resulting in a hypoxic environment that is injurious to certain species. Moreover, a semipermeable layer in the seed coat of most species restricts leakage and thus confounds the relationship of leakage with seed quality (Taylor et al., 1997). Furthermore, destructive methods involve biochemical techniques using staining procedures, complicated extraction protocols, and the use of potentially harmful reagents.

Therefore, new methods that allow insight into cellular integrity and function should provide a biophysical basis for predicting seed quality. The method should be rapid (conducted within a 24-h period) and not be biased by the seed coat or seed-covering tissues. Furthermore, the test should be nondestructive and noninvasive so that tested seeds may be used for other studies or purposes (Repo et al., 2002). Here are some of the nondestructive tests for measuring seed viability.

2.2.1 X-RAY ANALYSIS

The X-ray test has been routinely used to analyze seeds from different plant species. The test consists of a radiographic analysis of internal seed structures to detect any damage or abnormality that would restrict the seed germination. This technique does not require previous seed treatment and the low radiation level that is absorbed by the seed neither induces genetic mutations nor affects the germination performance. Besides, it is a precise, quick, easy-to-perform, and nondestructive method, generating additional information about seed viability (Bino et al., 1993).

The objective of the inclusion of the X-ray test in the Rules of Seed Analysis (ISTA, 1993) was to complement the information given by the germination test. The seed radiography allows the visualization of mechanical injuries, damage caused by insects, cracks, or fractures caused by pre- and postharvest handling (ISTA, 1993). Moreover, it allows the detection of embryo abnormalities, as well as the determination of their development stage. The use of X-ray test is increasing and it is useful at different stages of the seed production and utilization.

2.2.2 ELECTRICAL IMPEDANCE SPECTROSCOPY (EIS)

The method, electrical impedance spectroscopy (EIS), was introduced to study seed viability nondestructively. The EIS offers several advantages in comparison with conventional or physiological methods of assessing seed viability. The method generates a complete spectrum providing objective data to calculate a number of useful parameters. The technique is rapid and the actual measurement requires only minutes to complete. The entire procedure of preparing seeds and performing the EIS analysis can be conducted in less than 24 h. Seed moisture content has a strong effect on EIS parameters of seeds, and that the electrical properties of seeds change upon loss of seed viability.

2.2.3 MULTISPECTRAL IMAGING

Multispectral imaging is an emerging nondestructive technology in seed science, which integrates the conventional vision and spectroscopy technique to attain both spatial and spectral information from the target objects simultaneously. Multispectral imaging requires no sample pretreatments, making it more suited for process monitoring and quality control. This technique also has a great potential to measure the multiple components by reflection from both visual and near-infrared wavelengths at the same time for seed quality assurance. Vision and spectral technologies have shown promising results in different aspects of determining seed quality features such as fungal infection (Liu et al., 2014) and seed purity (Kong et al., 2013).

2.2.4 MICROOPTRODE TECHNIQUE (MOT)

A real-time and noninvasive microoptrode technique (MOT) was developed to measure seed viability in a quick and noninvasive manner by measuring the oxygen influxes of intact seeds, approximately 10 s to screen one seed. MOT is a highly sensitive and selective technique to measure oxygen concentrations and fluxes on the cell surface. MOT allows for the detection of single seeds and provides a novel method to compare the difference of activating one seed by oxygen or ion fluxes. The MOT is an ultrasensitive tool with high temporal and spatial resolution for detecting the physiological activity of live cells/tissues. Use of this methodology will provide novel insights into seed science research. In our routine work, we successfully employ MOT to quantify seed viability, by measuring surface oxygen influxes of intact seeds (Xia xin et al., 2013).

2.2.5 INFRARED THERMOGRAPHY (IRT)

Infrared thermography (IRT), a fully noninvasive technique, can discern the thermal profiles of highly viable, aged, and dead seeds upon water uptake. It is also useful to explore the biochemical basis of varying thermogenic activities in seeds of differential viability, and to develop an algorithm that predicts seed viability. IRT can link biochemical and biophysical parameters with developmental changes and visualize them noninvasively. IRT can also visualize in real time the earliest physicochemical events in seed germination and diagnose seed viability before radicle emergence, after which seeds can still be redried and restored (Llse Kraner et al., 2010).

2.2.6 SEED VIABILITY MEASUREMENT USING RESAZURIN REAGENT

A simple, quick, and nondestructive test method has been developed for determining Brassicaceae seed viability with single seed using resazurin reagent which was made by resazurin and yeast mixture. The color of the resazurin reagent will change from blue to pink or colorless when the aged seeds are soaked in the resazurin reagent solution for 4 h at 35°C. Seed soaking system also developed using 96-well plate and absorbance of the resazurin reagent to be measured at 570 nm with a multiplate reader. A model equation has been developed for predicting germination percentage of the seeds by the color fractions of the soaked resazurin reagent from the intact seed lots. The equation showed high prediction accuracy of 98.2%. This method was reported to be very quick and simple to use with a high accuracy and demonstrated with Brassicaceae seeds nondestructively (Tai Gi Min and Woo Sik Kang, 2011).

2.2.7 COMPUTERIZED SEED IMAGING

In the last two decades, the advent of computer-aided data acquisition by video camera, coupled with image processing and analysis, has allowed to capture time-lapse image sequences and to quantify several morphological features, necessary for germination and vigor testing.

The declining cost and increasing speed and capability of computer hardware of image processing and its integration with controlled environmental condition systems have made computer vision more attractive for use in automatic inspection of crop seeds. New algorithms and hardware architectures have been developed, and the availability of appropriate image analysis software tools suggests that the use of machine vision systems is becoming convenient in a seed biology laboratory.

The speed of operation of a machine vision system must allow rapid image processing and recording of measurements. Data may be further processed statistically and displayed graphically, and a database may be developed to integrate image analysis data with taxonomic and biomorphological features of plant species. So this integrated system can represent a new approach to understand seed biology and quality, and it includes operative system modeling and automation, digital imaging, data collection, and integration with those obtained from standard seed quality tests (ISTA rules, 2005), and new experimentation and hypothesis. The ultimate goal is not the management of system data, but their use for the development of mathematical models to describe and predict seed germination and quality as well as to extend seed analyst ability to control operations automatically.

SECTION III

CHAPTER

Plant growth measurements

3

Although crop species vary enormously in their assimilatory abilities, inherent growth in the plant still exerts powerful effects over their general performance. Physical inputs sustain growth, but biological regulation dictates the pattern of their utilization and ultimate expression. For individual cells growth is potentially unlimited and begins as an exponential pattern. However, mutual interactions with an individual impose limitation on growth. The actual growth curve falls away in a sigmoid manner. To understand the nature of growth regulation at a whole-plant level and appreciate the interactions between plants and their environment, it needs detailed growth measurements than simply final yield. Hence, different growth measurements useful in characterizing genotype capabilities are given in this chapter.

3.1 MEASUREMENT OF GROWTH

Growth can be measured by a variety of parameters as follows:

1. *Fresh weight:* Determination of fresh weight is an easy and convenient method of measuring growth. For measuring fresh weight, the entire plant is harvested, cleaned for dirt particles if any, and then weighed. Generally fresh weight is not considered, due to variable tissue water content.
2. *Dry weight:* The dry weights of plant organs are usually obtained by drying the materials for 21–48 h at 70–80°C and then by weighing it. This is the most useful and reliable measurement.
3. *Length:* Measurement of length is a suitable indication of growth for those organs that grow in one direction with almost uniform diameter such as roots and shoots. The shoot length (plant height) can be measured by a scale from ground to top growing point and expressed in centimeter. The advantage of measuring length is that it can be done on the same organ over a period of time without destroying it.
4. *Area:* It is used for measuring plant organs such as leaf. The area can be measured by a graph paper or by a photoelectric device (digital leaf area meter).

3.2 MEASUREMENT OF BELOW GROUND BIOMASS

Measurement of root biomass is equally important as the shoot biomass. By measuring the shoot biomass alone it would be difficult to know whether the apparent increase in above ground productivity is due to photosynthetic gain or simply because of redistribution of matter from the underground root.

Phenotyping Crop Plants for Physiological and Biochemical Traits. http://dx.doi.org/10.1016/B978-0-12-804073-7.00003-X

The entire process of root biomass study is accomplished in the following steps:

1. *Extraction of Root*
Root samples are taken from the center of the plot marked for shoot sample studies. Care should be taken to dig entire root system if experimenting under field conditions. For root studies growing crops on pipes or raised soil beds are preferred.
2. *Washing*
Roots should be completely washed off the clay, silt, sand, and organic matter. A simple root washing machine consists of a long cylinder (area = 1000 m^3) centrally fitted with a plunger having a perforated circular base. The plunger moves vertically and disperses the soil sample. Roots and organic matter are then decanted off.
A vortex root washer may also be used for this purpose. In the vortex washer water flows with speed and through the outflow roots and organic matter falls on the sieve band. Sand and larger particles fall to the base of the washer while clay and silt are passed through the sieve.
Iron sulfide deposits on roots (in case of waterlogged soil) can be removed by placing the washed root in continuously aerated water for 24–48 h.
3. *Removal of Dead Material*
The roots extracted and washed as above, should be separated from the decaying and dead matter (Hussey and Long, 1982). For this purpose, the extracted roots are put in solvents such as methanol or hydrogen peroxide. The living root material floats and dead root material sinks on the bottom of the container.
4. *Measurement of Dry Weight*
Measurement of dry weight of root is exactly the same as the above ground mass or the shoot. Determination of weight loss on ignition is particularly valuable for below ground biomass, since this eliminates contamination by inorganic soil mineral particles.

3.3 GROWTH ANALYSIS

Growth analysis is a mathematical expression of environmental effects on growth and development of crop plants. This is a useful tool in studying the complex interactions between the plant growth and the environment. Growth analysis in crop plants was first studied by British Scientists (Blackman, 1919; Briggs, Kidd, and West, 1920; William, 1964; Watson, 1952; Blackman, 1968). This analysis depends mainly on primary values (dry weights) and then can be easily obtained without great demand on modern laboratory equipment.

The basic principle that underlies in growth analysis depends on two values:

1. total dry weight of whole plant material per unit area of ground (W) and
2. total leaf area of the plant per unit area of ground (A)

According to the purpose of the data, leaf area and dry weights of component plant parts have to be collected at weekly, fortnightly, or monthly intervals. These data are to be used to calculate various indices and characteristics that describe the growth of plants and of their parts grown in different environments and the relationship between assimilatory apparatus and dry matter production. These indices and characteristics together are called growth parameters.

3.3.1 GROWTH CHARACTERISTICS—DEFINITION AND MATHEMATICAL FORMULAE

The following data are required to calculate different growth parameters to express the instantaneous values and mean values over a time interval.

1. *Crop Growth Rate (CGR)*: D.J. Watson coined the term crop growth rate. It is defined as the increase of dry matter in grams per unit area per unit time. The mean CGR over an interval of time t_1 and t_2 is usually calculated as shown in the following formula

$$\mathrm{CGR} = \frac{1}{P} \times \frac{W_2 - W_1}{t_2 - t_1} \left(\mathrm{g\,m^{-2}day^{-1}}\right)$$

   where W_1 and W_2 are the dry weights at two sampling times t_1 and t_2 respectively and P is the land area.
2. *Relative Growth Rate (RGR)*: The term RGR was coined by Blackman. It is defined as the rate of increase in dry matter per unit of dry matter already present. This is also referred to as "efficiency index" since the rate of growth is expressed as the rate of interest on the capital. It provides a valuable overall index of plant growth. The mean relative growth rate over a time interval is given as follows:

$$\mathrm{RGR} = \frac{\log_e W_2 - \log_e W_1}{t_2 - t_1} \left(\mathrm{g\,g^{-1}day^{-1}}\right)$$

   where $\log_e W_1$ and $\log_e W_2$ are the natural logs of dry weights at two sampling times t_1 and t_2, respectively.
3. *Net Assimilation Rate (NAR)*: The NAR is a measure of the amount of photosynthetic product going into plant material, that is, it is the estimate of net photosynthetic carbon assimilated by photosynthesis minus the carbon lost by respiration. The NAR can be determined by measuring plant dry weight and leaf area periodically during growth and is commonly reported as grams of dry weight increase per square centimeter of leaf surface per week. This is also called "unit leaf rate" because the assimilatory area includes only the active leaf area in measuring the rate of dry matter production.

   The mean NAR over a time interval from t_1 to t_2 is given by

$$\mathrm{NAR} = \frac{W_2 - W_1}{t_2 - t_1} \times \frac{\log_e A_2 - \log_e A_1}{A_2 - A_1} \left(\mathrm{g\,cm^{-2}wk^{-1}}\right)$$

   where W_2 and W_1 are plant dry weights at times t_1 and t_2, $\log_e A_2$ and $\log_e A_1$ are the natural logs of leaf areas A_1 and A_2 at times t_1 and t_2.

4. *Leaf Area Ratio (LAR):* The LAR is a measure of the proportion of the plant which is engaged in photosynthetic process. It gives the relative size of the assimilatory apparatus. It is also called a capacity factor. It is defined as the ratio between leaf area in square centimeters and total plant dry weight. It represents leafiness character of crop plants on area basis.

$$\text{LAR} = \frac{\text{Leaf area}}{\text{Leaf dry weight}}\left(\text{cm}^2\text{g}^{-1}\right)$$

5. *Leaf Weight Ratio (LWR):* It is one of the components of LAR and is defined as the ratio between grams of dry matter in leaves and total dry matter in plants. Since the numerator and denominator are on dry weight basis. LWR is dimensionless. It is the index of the plant on weight basis.

$$\text{LWR} = \frac{\text{Leaf weight}}{\text{Dry weight of plant}}$$

6. *Specific Leaf Area (SLA):* It is another component of LAR and defined as the ratio between leaf area in cm^2 and total leaf dry weight in grams. This is used as a measure of leaf density. The mean SLA can be calculated as

$$\text{SLA} = \frac{\text{Leaf area}}{\text{Leaf dry weight}}\left(\text{cm}^2\ \text{g}^{-1}\right)$$

7. *Specific Leaf Weight (SLW):* The reversal of SLA is called SLW. It is defined as the ratio between total leaf dry weight in grams and leaf area in cm^2. It indicates the relative thickness of the leaf of different genotypes.

$$\text{SLW} = \frac{\text{Leaf dry weight}}{\text{Leaf area}}\left(\text{g}^1\text{cm}^{-2}\right)$$

where W_L is the leaf dry weight and A is the leaf area.

8. *Leaf Area Index (LAI)*: D.J. Watson coined this term. It is defined as the functional leaf area over unit land area. It represents the leafiness in relation to land area. At an instant time (T) the LAI can be calculated as

$$\text{LAI} = \frac{\text{Total leaf area}}{\text{Land area}}$$

For maximum production of dry matter of most crops, LAI of 4–6 is usually necessary. The leaf area index at which the maximum CGR is recorded is called optimum leaf area index.

9. *Leaf Area Duration (LAD)*: It is usually expressed as a measure of leaf area integrated over a time period. Some takes into account both the magnitude of leaf area and its persistence in time. It represents the leafiness of the crop growing period. Thus, the unit measurement of LAD may be in days or weeks or months.

$$\text{LAD} = \frac{\text{LA}_1 + \text{LA}_2 (t_2 - t_1)}{2} (\text{cm}^2\text{d}^{-1})$$

$$\text{LAD (LAI basis)} = \frac{\text{LA}_1 + \text{LA}_2 (t_2 - t_1)}{2} (\text{cm}^2\text{d}^{-1})$$

where LA_1 and LA_2 are the leaf areas at two sampling times t_1 and t_2, respectively

10. *Harvest Index (HI):* Harvest index is the ratio of economic yield to the biological yield expressed in per cent. It represents the efficiency of photosynthate translocation to economic parts.

$$\text{HI} = \frac{\text{Economic yield}}{\text{Biological yield}} \times 100$$

Here, while calculating the biological yield we take only the above-ground parts into consideration.

CHAPTER

Photosynthetic rates

4

Photosynthesis is the cornerstone of physiological process and the basis of drymatter production in plants. Photosynthetic rate is an important parameter characterizing the photosynthetic capacity of the photosynthetic apparatus. It also reflects the efficiency because it is a determinant of light-use efficiency, biomass production, and crop yields. Photosynthesis is regulated through the control of green leaf area and stomatal conductance, which are variable according to genotype and its environmental interactions. Identification of high photosynthesizing genotypes under normal and stress conditions is essential to improve crop productivity. In this chapter, both direct and indirect methods of measuring photosynthetic efficiency of crop plants are described.

4.1 NET ASSIMILATION RATE (NAR)

NAR describes the net-production efficiency of the assimilatory apparatus. Net photosynthesis is equal to gross photosynthesis minus respiration. NAR is thus not quantifying the respiration since we get net photosynthesis, that is, net assimilation. Net assimilation rate can be computed as given later:

$$\mathrm{NAR} = \frac{W_2 - W_1}{t_2 - t_1} \times \frac{\log_e A_2 - \log_e A_1}{A_2 - A} \,\mathrm{g\,m^{-2}day^{-1}}$$

where W_2 and W_1 are plant dry weights at times t_1 and t_2, $\log_e A_2$ and $\log_e A_1$ are the natural logs of leaf areas A_1 and A_2 at times t_1 and t_2.

4.2 MEASURING THROUGH INFRARED GAS ANALYZER (IRGA)

Determination of WUE and associated physiological parameters by portable photosynthesis system/infrared gas analyzer (IRGA).

Principle: Infrared gas analyzers (IRGA) are used for the measurement of a wide range of hetero atomic gas molecules including CO_2, H_2O, NH_3, CO, SO_2, N_2O, NO, and gaseous hydrocarbons like CH_3. Hetero atomic molecules have characteristic absorption spectrum in the infrared region. Therefore, absorption of radiation by a specific hetero atomic molecule is directly proportional to its concentration in an air sample. Infrared gas analyzers measure the reduction in transmission of infrared

Phenotyping Crop Plants for Physiological and Biochemical Traits. http://dx.doi.org/10.1016/B978-0-12-804073-7.00004-1

wavebands caused by the presence of a gas between the radiation source and a detector. The reduction in transmission is a function of the concentration of the gas. The primary role of IRGA is to measure the CO_2 concentration. The IRGA is very sensitive to detect even a change of 1 ppm of CO_2.

A leaf or a plant is enclosed in an airtight chamber and the CO_2 fluxes are determined by measuring the CO_2 concentration changes in the chamber atmosphere. The major absorption peak of CO_2 is at 4.25 μm with secondary peaks at 2.66, 2.77, and 14.99 μm. Both water vapor and CO_2 molecules absorb IR radiation in the 2.7-μm range.

Procedure: The portable photosynthesis system is a portable IRGA and is designed to operate as an open system to measure the gas exchange parameters. It consists of separate IRGAs to measure CO_2 and H_2O vapor concentrations, an internal air supply unit and the necessary software for the computation of gas exchange parameters. Li 6400 uses four independent infrared gas analyzers, two each for CO_2 and H_2O. One pair of CO_2 and H_2O analyzers defined as reference measures the CO_2 and water vapor concentration in the ambient air that is sent into leaf chamber. Similarly second pair, the analysis chambers measure the CO_2 and water vapor concentrations in the air that is coming from the leaf chamber. The difference between the reference and the analysis IRGAs is computed. Deepa et al. (2012) measured physiological efficiency of greengram genotypes under moisture stress conditions in this laboratory.

A leaf is clamped to the leaf chamber. The leaf chamber is provided with suitable pads to clamp an area of 2.5 cm^2 under airtight conditions. Separate tubing is provided to send and withdraw air from the leaf chamber. These tubes are connected to either of the reference or analysis IRGA for the determination of gas concentrations.

A quantum sensor is placed inside the leaf chambers transparent cover to measure the actual light intensity in PAR range at the leaf surface. Blue and red light-emitting diode (LED) is fixed on top of the leaf chamber. The LEDs emit light in the PAR range and the intensity of which can be fixed and controlled at a required level. The light source is capable of providing the photosynthetically active radiation in the energy range of 0–2000 μmole m^{-2} s^{-1}.

A CO_2 cartridge normally carrying 8 g of pure CO_2 in a liquid form is used to get the requisite CO_2 concentration in the leaf chamber. The system mixes ambient air with the CO_2 to obtain the requisite concentration in the leaf chamber. The path of ambient air is provided with two scrubbers to remove moisture (drierite used as a desiccant) and CO_2 (soda lime to remove CO_2).

IRGA Working Procedure (LI 6400): Usage of IRGA (Fig. 4.1) equipment by students and scientists often found complicated. Here is the operation protocol for easy handling of the equipment both in greenhouse and field experiments.

Starting:

1. First charge the batteries 1 day prior to record data using IRGA.
2. Load the charged batteries first.
3. Connect the CO_2 tube to the inlet of the instrument.
4. All screws of this instrument must be in tight fitting.

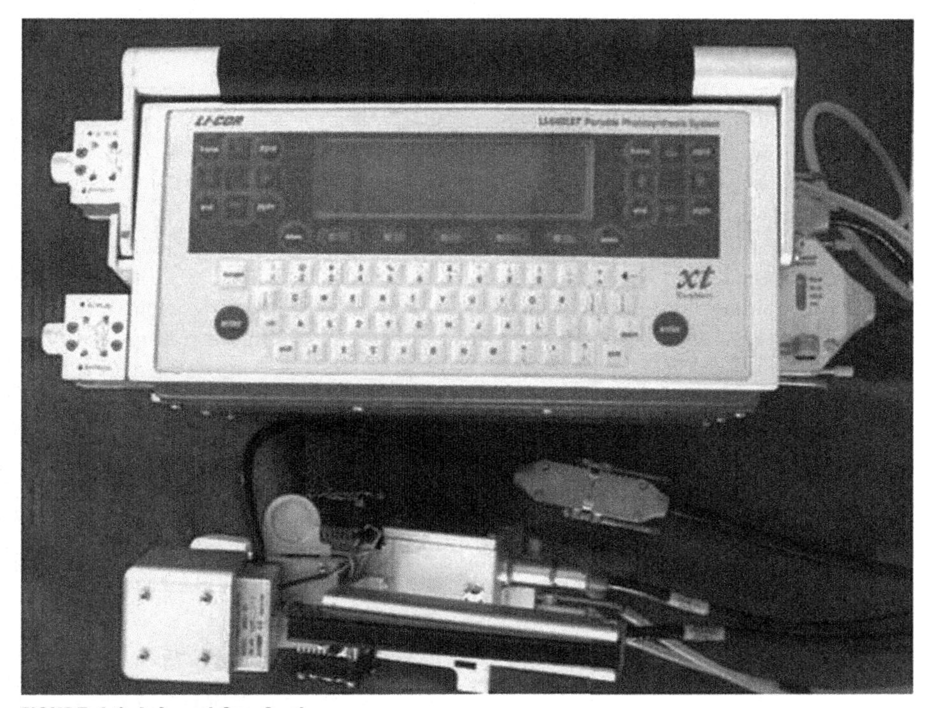

FIGURE 4.1 Infrared Gas Analyzer.

5. Connect the CO_2 tube in a proper way. Connect this tube very tightly; otherwise it shows leak (- ppm) in display.
6. The 2nd edge of this tube was kept in an empty thermocoal box and closed for uniform entry of air into the tube.
7. Switch "ON" the instrument.
8. Display shows -
   a. Welcome to loading open system.
   b. Starting net working.
   c. It shows fluorescence + WUE X m1 – press "enter."
   d. Is the chamber IRGA connected Y/S – Yes – press "Y."
9. Open the IRGA leaf chamber one time and close it.
10. Select "New measurements" press (F4).
11. In display select "Open log file" press (F1).
    a. Give file name and press "enter."
    b. Next—give sub file name and press "enter."
    c. Give date and press "enter."
12. Next—CO_2 matching.
    a. Select Match (F5).
    b. Wait up to we get equal values of reference CO_2 and sample CO_2.

c. If we need close matching press "Match IRGA" (F5) after that press "exit" (F1).

13. In Display set the rows – m, n, c and 9.
    a. If we want "m row"—press "m alphabet."
    b. If we want "n row"—press "n alphabet."
    c. If we want "c row"—press "c alphabet" it already exists.
    d. If we want "9 row"—press "9 number."
14. In this condition wait for 15–20 min for warming of instrument (before inserting the leaf in IRGA chamber).
15. Leaf should not fold in IRGA chamber. If leaf get fold it shows negative readings. Leaf should not have any moisture and dust before inserting leaf.
16. Insert the leaf in IRGA chamber.
    a. Give the "Dark pulse" (F3).
    b. Press "zero" getting "zero" row.
    c. Before going to next step, see the "*F*" value must be stable and df/dt value is <5.
    d. Select DOF_0F_m – (F3).
17. Select row no: 9: press "Actinic On" (F4).
18. Select row no: 8: press "Define actinic" (F3).
    a. It shows "Actinic Definition – press "enter."
    b. Type 1000 (PAR value 1000) press "enter."
19. Select "zero" row.
    a. Before going to next step, see them, "F" value must be stable and df/dt value is <5.
    b. Select DOFsF_0F_m – (F4).
20. If we want fluorescence value select "O" alphabet and note down the Fv'/Fm' value
21. Now note down the IRGA readings (photosynthetic rate, transpiration rate, stomatal conductance).
22. Before taking next reading "Actinic is in OFF" (F4). Do as above for taking every next reading.
23. Time taken for each reading is 10–20 min.
24. After taking of readings IRGA chamber must be in open condition (loose the screw).
25. Replace the fluorescence chamber foam (White foam) at the time of entire damage.

Shutdown:

1. In every shut down process "Actinic" must be in "Off" condition.
2. Press "Escape button."
3. Select "Utility menu – F5."
4. Coming down using down arrow.
5. Select "Sleep."
6. Give "Enter."

7. It shows – Ok to sleep Y/N.
   a. Press Yes – "Y" alphabet.
8. Switch off the system.
9. Disconnect the CO_2 tube.
10. Keep batteries for charging.

Parameters recorded from IRGA:

1. Photosynthetic rate (Photo): μmole CO_2 m^2 s^{-1}
2. Stomatal conductance (Cond): mole H_2O m^2 s^{-1}
3. Transpiration rate (Trmmol): m.mole H_2O m^2 s^{-1}
4. Intercellular CO_2 concentration (Ci): μmole CO_2 $mole^{-1}$
5. Chlorophyll Fluorescence (Fv′/Fm′ values)
   where Fv′ = variable fluorescence; Fm′ = maximum fluorescence.

4.3 RUBISCO ENZYME ACTIVITY

The enzyme Rubisco (ribulose bisphosphate carboxylase/oxygenase) has some special features of its own. It is the only enzyme which can catalyze carboxylation or oxygenation reaction depending upon the molecular concentration of CO_2 or O_2. As a carboxylase enzyme, it catalyzes combination of RuBP and CO_2 resulting in the formation of two molecules of 3-PGA. As an oxygenase, it plays the key role in the production of p-glycolate, the first intermediate in the photo-respiratory pathway. Rubisco constitutes more than 50% of soluble leaf protein. This indicates how important this enzyme into the plant.

In C_3 plants, Rubisco is located in the stroma of all chloroplasts. However, in C_4 plants it may be restricted to the chloroplasts of the bundle sheath cells.

There are eight active sites per molecule of Rubisco. The Michaelis constant (K_m) value of Rubisco for CO_2 is around 10–20 μM and for O_2 is around 200 μM. High K_m values reflect low affinity.

Rubisco is one of the most difficult enzymes to assay. This is because it is converted from an inactive to an active form by reaction with CO, and Mg^{++} and inactivation readily occurs in their absence. Another complication is that the extracted enzyme appears to be cold inactivated in certain plants such as wheat (Coombs et al., 1987). The amount of active Rubisco in a leaf is an important factor regulating the rate of photosynthetic carbon fixation (Servaites et al., 1984).

4.3.1 MEASUREMENT OF RUBISCO ACTIVITY

Principle: Assay of Rubisco is generally carried out by radiometry technique where radioactive CO_2 is used as a substrate and traces of radioactivity in the products are counted as a measure of enzyme activity.

Chemicals required:

1. Tris-hydroxymethyl amino methane
2. Hydrochloric acid (HCl)

3. Magnesium chloride ($MgCl_2$)
4. Dithiothreitol (DTT)
5. Acetic acid
6. Toluene
7. 2,5-Diphenyl oxazole
8. 1,4-bis-2-methyl, 5-phenyl oxazolyl-benzene

Extraction:

1. Collect the leaf sample on sunny days and bring it in a butter paper bag, kept in an ice box.
2. Weigh 0.2 g of freshly cut leaf pieces after blotting them dried.
3. Grind the sample in a prechilled pestle with mortar using 5 mL of ice cold grinding medium [the grinding medium consists of Tris-HCl buffer, 50 mM (pH 8.0), $MgCl_2$, 5 mM, and dithiothreitol (DTT), 10 mM].
4. Centrifuge the homogenate at 25,000 × g at 0–4°C for 5 min.
5. Take the supernatant, measure its volume, put it in a sample vial fitted with lid/cap, and keep it under ice.

Estimation:

1. Measure the initial activity at 25°C by injecting 50 μL of soluble leaf extract into an assay mixture containing 50 mM Tris-HCl (pH 8.0), 20 mM $MgCl_2$, 0.1% (w/v) BSA, 0.5 mM RuBP, 10 mM $NaH_{14}CO_3$ (1 μ Ci/assay) in a total volume of 0.5 mL (the initial enzyme activity is measured immediately after extraction).
2. Measure the total activity in a similar manner with an exception that 50 μL of soluble leaf extract and 350 μL of the assay mixture are incubated together at 25°C for 10 min in a shaking water bath before 100 μL of 2.5 mM RuBP is added.
3. Stop the reaction by adding 0.1 mL of 6 M acetic acid.
4. Dry tightly covered sample vials at 65°C.
5. To this, add 10 mL of cocktail [cocktail contains 2 g of PPO (2,5-diphenyloxazole) and 50 mg POPOP (1,4-bis-2-methyl, 5-phenyl oxazolyl-benzene) dissolved in 500 mL of toluene].
6. Determine acid stable (3-PGA) radio-activity in a liquid scintillation counter.
7. Run a blank similarly without RuBP (instead RuBP, add 0.05 mL of buffer).
8. Measure total soluble protein content in the crude enzyme extract by Bradford (1970) method.
9. Rubisco activity is expressed as [μmol CO_2 (g protein)$^{-1}$ s^{-1}].

As mentioned earlier, Rubisco must be activated fully to achieve maximal activity (Lorimer et al., 1976). According to Kobza and Seemann (1988) percent activation of Rubisco can be calculated using the following formula

Calculation:

$$\text{Activation}\,(\%) = (\text{Initial activity}/\text{Total activity}) \times 100$$

4.4 CHLOROPHYLL FLUORESCENCE RATIO (FV/FM VALUES)

Chlorophyll fluorescence is the most widely used technique in photosynthesis and plant stress research. Photochemical efficency of Photosystem (PS)-II is measured as chlorophyll fluorescence in terms of Fv/Fm ratio (variable fluorescence/maximum fluorescence). The extent of photoinhibition induced by any environmental stress can be rapidly assesed by measuring the maximum photochemical effiency of PSII. Chlorophyll fluorescence is measured directly by IRGA as explained earlier in this chapter.

CHAPTER

Drought tolerance traits

5

In agriculture, the term "drought" refers to a condition in which the amount of water available through rainfall and/or irrigation is insufficient to meet the transpiration needs of the crop. In general, a clear distinction should be made between traits that help plants to survive a severe drought stress and traits that mitigate yield losses in crops exposed to a mild or intermediate level of water stress. Ultimately, the knowledge contributed by any drought-related study should help address the yield issue either directly or indirectly.

An important prerequisite for the successful phenotyping of secondary traits is to identify key functional attributes that contribute to drought tolerance. Ideally, such identification is based on evidence that there is genetic variation for the trait in the crop of interest, on the trait being correlated with crop performance in drought environments, and on its having sufficient heritability to be used to make progress in a breeding program. Another criterion is that a trait has a clear-cut and rational explanation for its physiological function in drought tolerance. In this context, various reliable physiological traits for phenotyping drought tolerance in crop genotypes are described in this chapter.

5.1 WATER USE EFFICIENCY (WUE) TRAITS

WUE is one of the physiological traits that is associated with drought tolerance. It is the amount of dry matter produced to unit amount of water transpired. Increase in WUE in rainfed crops is the best method to improve yields. Crop yields under field conditions can be explained by a simple biological yield model by Passioura (1986).

$$\text{Grain yield} = T \times \text{WUE} \times \text{HI}$$

where T represents the total transpiration by the crop canopy;

WUE, the amount of biomass produced per unit amount of water transpired; and

HI, the ratio of biomass that is partitioned to economically important parts.

This relationship assumes greater importance when the water availability is limited. Under water limited conditions species or genotypic variations in biomass accumulation depend on T and WUE. The total transpiration is the cumulative effect of the average rate of water uptake, duration of transpiration and canopy cover. Along with transpiration, the WUE is yet another important parameter, which influences the biomass production. WUE is expressed as the ratio of biomass produced (g) to the amount of water transpired (kg) (Fischer and Turner, 1978).

Phenotyping Crop Plants for Physiological and Biochemical Traits. http://dx.doi.org/10.1016/B978-0-12-804073-7.00005-3

Variability in WUE is mainly determined by two methods:

1. Direct method
2. Indirect method

1. Direct methods
   WUE is quantified directly by two methods:
   a. At single leaf level
   b. At whole plant level
      - *At single leaf level*: WUE at single leaf level is quantified by adopting gas exchange approach.
        WUE at single leaf level is the ratio of carbon assimilation rate (A) to transpiration (T) and that can be well described by the following equation:

$$\mathrm{WUE} = (A / T)$$

        Transpiration rate is determined by the intrinsic stomatal conductance and the existing leaf to air vapor pressure difference (v). If the plants are grown under very similar environmental condition, it can be expected that the leaf to air vapor pressure difference will be similar and hence, the major factor that determines transpiration would be the intrinsic stomatal conductance (g_s).
        Therefore, the equation becomes WUE = (A/g_s).
        Both A and g_s are gas exchange parameters and can be determined by the portable photosynthesis system/IRGA.
      - *At whole plant level*: At whole level, WUE determined in small pots by adopting gravimetric approach.

Determination of WUE and associated physiological parameters by gravimetric approach: Determination of WUE which can be achieved by using smaller pots, determining water loss from each pot gravimetrically on a regular basis (eg, daily) and replacing part of the transpired water to control the rate of soil dry-down. This is called gravimetric approach. Latha and Reddy (2005) determined WUE in groundnut genotypes by gravimetric method.

Principle: This technique involves the accurate determination of the total water transpired by a plant over a specific period of crop growth and the total biomass the plant accumulated over the same period. Plant tissues were exposed to drought stress at whole plant level for determination of various physiological parameters.

Materials required: Pots having handles (battery pots), mobile weighing balance, rain out shelter.

Procedure: At the whole plant level, water use and WUE can be determined during a period of 30 days at an active growth phase period. Although this is the best period for determining genetic variability, the duration and stage of determination can vary, depending upon the objectives of the investigation. This approach relies on the accurate determination of water content in the soil by weighing the containers at regular intervals, at least once every day. Since transpiration is being monitored, it is essential to ensure that there is no addition of water to the containers from any external source

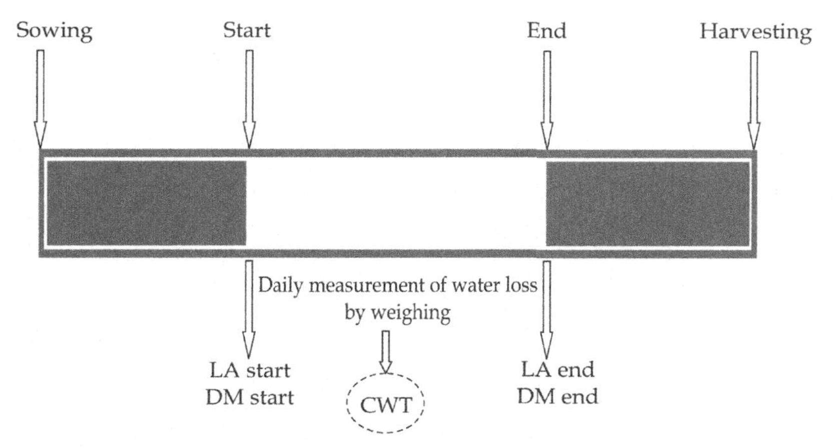

(CWT = cumulative water transpired; DM = dry matter; LA = leaf area.)

FIGURE 5.1 Flowchart Indicating the Sequence of Events for Determining WUE and Other Associated Parameters by a Gravimetric Approach (as Shown in the Figure).

such as rain. Thus, it is preferable to lay out the experiment under a mobile rainout shelter. This structure can be moved over the area where containers are placed during the night and/or during any rain episodes (Udayakumar et al., 1998).

The protocol is shown as follows (Fig. 5.1):

- In this approach plants are grown in suitable containers filled with a rooting mixture consisting of red sandy loam soil and farmyard manure (3:1 by volume) or any other substrate suitable for that particular crop.
- The containers are weighed once or twice daily and the difference in weight on subsequent days is corrected by adding an exact amount of water.
- Initially, it is necessary to assess the weight of the container when empty (W_E) and then after filling with either dry soil or a specific rooting mixture (W_{ES}).
- The amount of dry soil (W_S) in each container would then be calculated from:

$$W_S = W_{ES} - W_E$$

- A hanging load cell balance mounted on a mobile gantry (Fig. 5.2) with a provision for movement on rails on either side of the rainout shelter can be used to access and weigh each container.
- At the start of the experiment, the soil should be brought to 100% field capacity by adding the appropriate volume of water, which can be determined by considering the bulk density of the soil and its water holding capacity.
- At this stage, the drainage holes of the containers are closed to prevent the added water from draining out.
- The soil surface should be covered with plastic or any other suitable mulching material to minimize water loss due to surface evaporation. These arrangements ensure that the majority of the added water is available for transpiration only.

FIGURE 5.2 A Hanging Load Cell Balance Mounted on a Mobile Gantry to Weigh the Pots.

- A control set of four to five containers without plants should be maintained with exactly the same soil, water, and mulching to provide an accurate measure of surface evaporation.
- The pots with plants and the controls are weighed at least once daily and sufficient water added to bring the soil back to 100% field capacity.
- The water added daily over the entire experimental period is summed to arrive at the cumulative water added (CWAP) to the pots with plants.
- The total transpiration can be determined by subtracting the cumulative water added to the control containers (CWAC) from CWAP. Thus:

$$\text{Cumulative water transpired (CWT)} = \Sigma(\text{CWAP}) - \Sigma(\text{CWAC})$$

- At the start of the experiment, total biomass and leaf area are determined in a set of three containers.
- The soil is carefully washed with a jet of water to remove the roots, and the plant parts (leaves, stem, and roots) are separately oven dried at 70°C for 3 days.
- Biomass and leaf area are recorded again for the plants in the remaining containers at the end of the experiment.

- Assuming linear growth during the experimental period, WUE and other physiological parameters are calculated as follows:

$$WUE\,(g\,kg^{-1}) = (DM_{end} - DM_{start}) / CWT$$

where DM_{end} and DM_{start} are the total dry matter (g pot^{-1}) measured at the end and start of the experiment, respectively. Then, leaf area duration (LAD) can be calculated as

$$LAD\,(cm^2\,days) = [(LA_{end} + LA_{start}) / 2] \times \text{duration of experiment (days)}$$

where LA_{end} and LA_{start} are the leaf area (in cm^2 $plant^{-1}$) measured at the end and start of the experiment, respectively. The net assimilation rate (NAR) and the mean transpiration rate (MTR) are time-averaged measures of photosynthetic rate and transpiration rate, and are calculated as follows:

$$NAR\,(g\,cm^{-2}\,day^{-1}) = (DM_{end} - DM_{start}) / LAD$$

$$MTR\,(mL\,cm^{-2}\,day^{-1}) = CWT / LAD$$

The novelty of the gravimetric approach is that, besides determining WUE, a few important physiological traits such as NAR, MTR, and LAD can also be calculated.

2. Indirect methods
   a. *Specific leaf area*: The specific leaf area (SLA) is the ratio of leaf area to leaf dry weight, and is an indirect measure of leaf expansion. Higher SLA values represent a larger surface area for transpiration, Hence, SLA and WUE would be inversely related. Studies by Latha and Reddy (2007) and Rao and Wright (1994) demonstrated a positive correlation between SLA and $\Delta^{13}C$ ($r = 0.90$–0.93) in groundnuts, and a negative relationship between SLA and WUE, suggesting that SLA can be used as an alternative for rapid estimation of genetic variability in WUE among groundnut genotypes. Although a close correlation has been established between SLA and $\Delta^{13}C$ (and thus with WUE) in controlled experiments, the strength of correlation varied ($r = 0.71$–0.94) over a range of groundnut genotypes and environments (Wright et al., 1994). It can therefore be inferred that SLA might be influenced by factors such as time of sampling and leaf age (Rao et al., 1995; 2001; Wright and Hammer, 1994), as well as the accuracy of measurement.

Procedure

1. The second or third completely expanded leaf from the apex of the main stem is selected.
2. The actual leaf area is recorded immediately using a leaf area meter (Fig. 5.3).
3. The leaf is then dried in a hot air oven at 70°C for 3 days.
4. Measure the dry weight of the leaf accurately using a sensitive balance.

FIGURE 5.3 Leaf Area Meter (LI-3100C).

5. SLA is calculated as follows:

$$SLA = \frac{\text{Leaf area}}{\text{Leaf dry weight}} \, cm^2 g^{-1}$$

b. *SPAD Chlorophyll Meter Reading (SCMR)*
The light absorbance and/or transmittance characteristics of a leaf can be exploited to determine the leaf chlorophyll content (Balasubramanian et al., 2000; Takebe et al., 1990). Simple handheld instruments are now available that measure a unit less value which is directly related to the chlorophyll content. These instruments determine the light attenuation at 430 and 750 nm. One such instrument is the Soil Plant analytical Development (SPAD) chlorophyll meter (Fig. 5.4) and the unit less number displayed is referred to as the SPAD chlorophyll meter reading (SCMR). Sudhakar et al. (2006) reported a significant negative correlation between SCMR and SLA ($r = 0.73$) in blackgram and greengram genotypes. A significant positive relationship between SCMR and chlorophyll content has been reported in many crop species including groundnuts (Rao et al., 2001; Sheshshayee et al., 2006). These authors also demonstrated a strong correlation of SCMR with SLA and SLN, corroborating earlier reports (Chapman and Barreto, 1997; Dwyer et al., 1995). Thus, SCMR is being used as a simple alternative technique to estimate differences in WUE, at least as an initial screening.

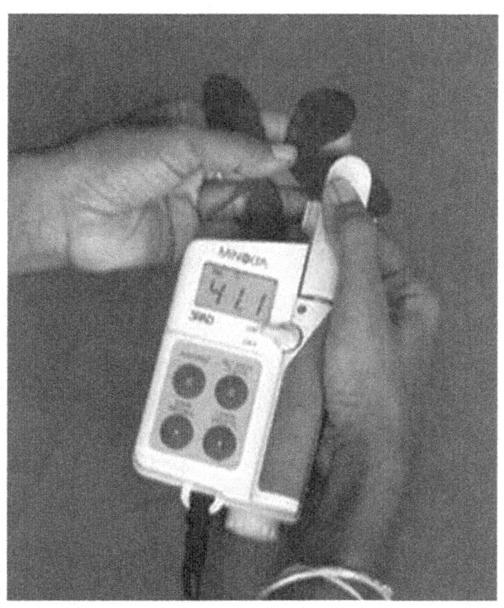

FIGURE 5.4 SPAD Chlorophyll Meter.

Procedure:

1. Normally, the second or third completely expanded leaf from the apex is chosen.
2. The leaf lamina avoiding the mid-rib portion is clamped into the sensor head of the SPAD meter.
3. A gentle press is given to record the SCMR value and the average of 30 measurements can be recorded by average nob. In the case of crops, where leaf tri or tetrafoliate, each of the leaflets of the second leaf/third leaf is used to take the readings.
4. The SPAD readings are more stable under natural light between 10.00 and 16.00 h.

   c. *Stable isotope ratio*:

   It is a powerful time-averaged option for estimating physiological traits. The isotope composition of carbon and oxygen provides very useful time-averaged information on several physiological traits such as WUE, transpiration rate, carboxylation efficiency, and root traits. Furthermore, determination of stable isotope ratios in a continuous flow mass spectrometer is high-throughput. Hence, stable isotope ratios provide a very powerful option to estimate physiological traits in a large number of accessions and breeding lines.

   Being accurate and high-throughput, stable isotope ratios of carbon and oxygen provide an attractive option for phenotyping a large number of germplasm accessions, as well as mapping populations derived by crossing contrasting genotypes. Furthermore, analysis of carbon and oxygen isotope composition offers proof of the use of stable isotope signatures as powerful surrogates for accurate phenotyping of these complex physiological traits in large numbers of accessions (Impa et al., 2003).

5.1.1 CARBON ISOTOPE DISCRIMINATION

Carbon isotope discrimination ($\Delta^{13}C$) measures the ratio of stable carbon isotopes ($^{13}C/^{12}C$) in the plant dry matter compared with the ratio in the atmosphere. Because of differences in leaf anatomy and the mechanisms of carbon fixation in species with the C3 or C4 pathway, studies on $\Delta^{13}C$ have wider implications for C3 species where the variation in $\Delta^{13}C$ is larger than in C4 species and has a greater impact on crop yield. Commonly, $\Delta^{13}C$ is negatively associated with WUE over the period of dry mass accumulation (Araus et al., 2002; Condon et al., 2004; Latha and Reddy, 2007).

During photosynthesis, plants discriminate against the heavy isotope of carbon (^{13}C), resulting in the depletion of ^{13}C content in biomass compared with atmospheric air (O'Leary, 1981). The extent of this depletion in the carbon isotope ratio ($^{13}C/^{12}C$), called "carbon isotope discrimination" ($\Delta^{13}C$), has been shown to be related to the ratio of the partial pressures of CO_2 inside the leaf to that in ambient air (Pi/Pa), as follows (Farquhar et al., 1989a; Hubick et al., 1986):

$$\Delta^{13}C = a + (b - a)Pi/Pa$$

where a and b are isotope fractionations that occur during diffusion through stomata and carboxylation by Rubisco, respectively (Farquhar et al., 1982; Hubick and Farquhar, 1989; O'Leary, 1981).

Stomatal diffusion and carboxylation processes also regulate transpiration and photosynthesis, and hence WUE. The theory linking $\Delta^{13}C$ and WUE has been well studied and the physiological basis for such a relationship is also well understood (Farquhar et al., 1989a; 1989b). Several field and container experiments have validated the association between WUE and $\Delta^{13}C$, suggesting that $\Delta^{13}C$ is a powerful surrogate for WUE in both annual and perennial plants.

5.1.2 DETERMINATION OF STABLE CARBON ISOTOPES USING ISOTOPE RATIO MASS SPECTROMETER (IRMS)

An isotope ratio mass spectrometer (IRMS) interfaced with a suitable combustion system is used for the determination of stable isotope ratios on a continuous flow basis. The technique involves the conversion of solid organic matter into constituent gas species and their introduction into the ion source, along with a helium carrier stream, for determination of isotope ratios. Carbon isotope ratios are determined by combusting organic matter in specialized reactor filled with suitable catalysts for quantitative combustion. In case of oxygen isotope ratios, samples are pyrolysed at high temperatures (1400°C) in the complete absence of oxygen. While CO_2 gas is introduced into the ion source for carbon isotope ratio determination, it is the CO gas that is produced during pyrolysis that is introduced for oxygen isotope ratio measurements. The reader is advised to consult other reviews for a greater understanding of isotope ratio mass spectrometry (Burlingame et al., 1998; Ehleringer and Osmond, 1988).

5.1.3 PROTOCOL FOR CARBON ISOTOPE DISCRIMINATION IN LEAF BIOMASS

Collection of soil sample:

1. The choice of the sample and its preparation are the most important aspects determining the accuracy of stable isotope ratio measurement. The sample to be chosen for isotopic analysis depends on the objectives of the investigation.
2. If the objective is to look for ontogenic differences, leaf samples of specific age should be collected.
3. If the objective is to study physiological processes like WUE and transpiration rate during a treatment period, then the samples can be collected during the treatment period of the experiment (typically, the third or fourth leaf from the apex can be harvested).
4. A composite leaf sample is also taken. In this approach, mature leaves representing the entire canopy can be harvested and dried.
5. Care must be taken to avoid collecting very young or senescing leaves.

Preparation of sample:

1. The combustion and pyrolysis systems require extremely small quantities of the sample, typically in the range 0.6– 1.0 mg.
2. Any small amounts of contamination can significantly alter the isotopic signatures measured. Therefore, while preparing the samples, it is extremely important to take utmost care.
3. The samples are dried to complete dryness at a temperature of 70°C for a period of 3 days in a hot air oven.
4. They are then powdered and completely homogenized using a pestle and mortar or a ball mill, taking care to wipe the pestle and mortar or the ball mill with a clean nonlint tissue paper dipped in acetone to ensure that there is no sample contamination.
5. The powdered samples of organic matter are then placed in capped glass or plastic vials and stored in a dry place until use.

Although stable isotope ratios are accurate surrogates, progress in breeding for physiological traits can be hindered severely by the cost of a mass spectrometer and limited accessibility. Thus, simpler yet accurate alternatives are essential, at least as initial screening strategies.

5.2 ROOT TRAITS

Phenotyping for high root traits: Variability in root traits is mainly determined by two methods:

1. Direct method
2. Indirect method

1. *Direct method:*
   a. *Raised soil bed method (root structure):*
   Measurement of root traits: Determination of genetic variability in root traits represents the most difficult challenge in crop improvement programmes. Despite the undeniable importance of root traits in better water mining, progress in breeding for these traits has been extremely slow, owing to the difficulty of determining the below ground biomass. Several techniques have been developed ranging from hydroponics to growing plants in pots and pipes (Venuprasad et al., 2002). Such approaches have the inherent disadvantage that root growth is constrained by the space available. These disadvantages can largely be overcome by raising plants in specially constructed "root structures."

Procedure:
Construction of root structures:
- Although various dimensions can be adopted, the most suitable for most of agriculture crops would be constructing 5 ft tall, 10 ft wide, and 60 ft long structures using cement bricks (use less cement, so as to remove brick by brick at harvest).
- An additional 5 ft tall wall can be built in the middle of the structure to make two halves, each 5 ft wide.
- Then fill this structure with soil having good texture and allow it for one monsoon season to allow natural compactness comparable to soil of experiment farm.
- Draw soil samples from these raised soil beds to quantify soil properties, viz., bulk density, particle density, water-holding capacity, and porosity.
- If these soil properties are comparable to the natural soil of experiment farm, sowing can be taken up.

Experimenting on root structures
- Seeds can be planted in rows and an exact plant population can be maintained.
- At the end of the experiment, the brick walls along the sides can be dismantled with care and the wash the soil using a strong jet of water.
- The roots must be separated carefully from soil particles and record various parameters such as root length, number of primary and secondary roots, volume, etc.
- The roots can then be separated from the shoots and oven dried to measure root biomass.

FIGURE 5.5 Specially Designed Root Structure and Variability of Root Traits Among Different Groundnut Genotypes.

Except for the fact that the plants are grown in raised structures, this approach provides an option for determining genetic variability in several root traits under conditions that are almost natural. (Root structure and variability as root traits are shown in Fig. 5.5.)

2. *Indirect method:*
   a. *Stable oxygen isotope:*
   The use of stable oxygen isotopes has generated considerable interest in plant carbon and water relations in recent years. It has been fairly well documented that during evaporation of water from lakes and oceans, water gets enriched with the heavy isotope of oxygen (^{18}O) because the water molecules containing lighter isotopes diffuse relatively faster and have a higher vapor pressure than those with heavy oxygen isotope (Craig and Gordon, 1965). Being an evaporative process, transpiration would result in a similar enrichment in heavy oxygen ($\Delta^{18}O$).

   Accordingly, several reports have demonstrated that the leaf water is indeed enriched with ^{18}O compared with the source or xylem water (Barbour and Farquhar, 2000; Flanagan, 1993). However, the relationship between $\Delta^{18}O$ and stomatal conductance had remained equivocal. Bindu Madhava et al. (1999) and Sheshshayee et al. (2005) demonstrated that oxygen isotope enrichment is positively related to stomatal conductance, at a given constant vapor pressure deficit (VPD). A mathematical explanation for this relationship was also provided by Sheshshayee et al. (2010). Since the ^{18}O signature of the leaf water is progressively imprinted in organic molecules (Sternberg et al., 1986), the $\Delta^{18}O$ of the leaf biomass is, therefore, an integrated value of the intrinsic stomatal conductance, and it also accounts for the diurnal and seasonal influence of weather parameters on transpiration rate.

Protocol for quantification of ^{18}O composition in leaf biomass

1. The oxygen isotopic composition of biomass is generally determined by quantitative pyrolysis of the sample at high temperature in the complete absence of oxygen.
2. The dried leaf powder (0.8–1.2 mg) was taken in silver capsules and placed in sample carousel of the autosampler.
3. The autosampler drops the sample at precisely designated times sequentially into the pyrolysis column.
4. The pyrolysis column contains a graphite tube filled with glassy carbon catalyst. The graphite tube is placed in a ceramic column heated to 1400°C.
5. The pyrolysis of organic matter resulted in the production of carbon monoxide (CO) and di-nitrogen (N_2). CO_2 is generally <5%.
6. These gases are swept by a helium (He) carrier gas (purity >99.996 %), into a GC column and then into the ion source of IRMS.

Since the mass-to-charge ratio (*m/z*) of CO and N_2 is the same, it is essential to separate these two gases before introducing into the ion source for isotopic ratio discrimination. These gases were passed through a GC column containing 5 Å molecular sieve heated to 90°C. The CO travels slower than N_2 gas through molecular sieve; hence the two gas species can be quantatively separated. The *m/z* ratio for 28 and 30 masses corresponding to $C\,^{16}O$ and $C\,^{18}O$ respectively is determined by the IRMS. An appropriate standard (Craig-corrected against SMOW) was also introduced in the run to determine the accuracy of mass detection and standard deviation of the run. The ^{18}O enrichment over and the source was computed as follows:

$$\Delta^{18}O(0/00) = \delta^{18}O_{\text{sample}} - \delta^{18}O_{\text{water}}$$

$\delta^{18}O$ was determined by CO_2 equilibration technique using the gas bench coupled to the IRMS. The $\delta^{18}O$ of irrigation water is found to be $-3.73^{0}/_{00.}$

For carbon isotope discrimination and oxygen isotope studies related to WUE measurements, results can be obtained on a paid service basis from Department of Crop Physiology, University of Agricultural Sciences, Bangalore where national facility is created for IRMS.

CHAPTER

Other drought-tolerant traits

6

6.1 RELATIVE WATER CONTENT (RWC)

RWC is a physiological parameter to assess the drought-tolerant capacity and photosynthesis efficiency of the plants. It is useful index for screening drought-tolerant cultivars. Turgidity is the function of available water and it invariably decides the magnitude of physiological processes that are occurring in leaf tissue. The RWC decides the magnitude of the physiological processes and plant metabolism.

Procedure:

Relative water content (RWC) of leaf discs is quantified according to Barrs and Weatherlay (1962).

1. A physiologically functional leaf (third from the top) is taken and leaf punches of same diameter numbering 10–20 are taken or whole of known number can be used.
2. Record the fresh weight.
3. The leaf discs/leaves are floated in 10 mL of water for 6 h and allowed to gain turgidity.
4. Take out the leaf discs/leaves and remove the water particles on the sample with a blotting paper.
5. Turgid weights are recorded and dried in hot air oven at 70°C.
6. Record the dry weight of the leaf discs/leaf samples.
7. RWC is estimated and expressed in percentage using the following formula:

RWC (%) = [(Fresh weight – Dry weight) ÷ (Turgid weight – Dry weight)] ×100
Saturation water deficit (SWD) = 100-RWC

6.2 CHLOROPHYLL STABILITY INDEX (CSI)

CSI is a measure of integrity of membrane or heat stability of chlorophyll under stress condition. The first change observed on plants suffering from drought is the wilting of leaves and the gradual fading of the green color. When color change reaches some critical point, recovery is no longer possible. Membrane permeability is lost under stress conditions which lead to protein denaturation of pigment and oxidation of pigments resulting in lesser capacity to absorb light. The CSI is a function of temperature and this property of chlorophyll is found to correlate with drought resistance.

Phenotyping Crop Plants for Physiological and Biochemical Traits. http://dx.doi.org/10.1016/B978-0-12-804073-7.00006-5

Principle: The high CSI value indicates the better availability of chlorophyll in the plant and helps the plant to withstand stress. This leads to increased photosynthetic rate, more dry matter production, and higher productivity. It was calculated as the difference in light transmission percentage between treated (kept at 65°C for an hour) and untreated leaf samples (kept at ambient temperature for 1 h).

Chemicals required: Dimethyl sulfoxide (DMSO).

Extraction:

- *Control*: Place 0.1 g of fresh plant leaf material in a test tube and add 10 mL of DMSO.
- *Treated:* Place 0.1 g of fresh plant leaf material in a test tube and heat it in a water bath for 1 h at 65°C and after heating, add 10 mL of DMSO.
- Keep both treated and untreated samples for overnight.

Estimation: Chlorophyll extracted into DMSO solution was collected from test tubes and concentration of chlorophyll a, b and total chlorophyll were quantified in both treated and untreated samples by reading the optical density at 663 and 645 nm. Calculate chlorophyll a, b and total chlorophyll using the formula given later.

Calculation: Calculate the amount of chlorophyll present in the extract (treated and control) in milligram chlorophyll per gram tissue using the following equations:

$$\text{mg chlorophyll a/g tissue} = 12.7(A\ 663) - 2.69(A\ 645) \times \frac{V}{1000 \times W}$$

$$\text{mg chlorophyll b/g tissue} = 22.9(A\ 645) - 4.68(A\ 663) \times \frac{V}{1000 \times W}$$

$$\text{mg total chlorophyll/g tissue} = 20.2(A645) + 8.02(A663) \times \frac{V}{1000 \times W}$$

where A = absorbance at specific wavelengths,
V = final volume of chlorophyll extract in DMSO
W = fresh weigh of tissue extracted.

Now calculate the chlorophyll stability index of the leaf material using the following formula

$$\text{CSI} = \frac{\text{Total chlorophyll of the heated sample}}{\text{Total chlorophyll of the untreated sample}} \times 100$$

6.3 SPECIFIC LEAF NITROGEN (SLN)

Variations in WUE are brought about either by stomatal factors or by the capacity of chloroplasts to fix carbon. The latter depends significantly on biochemical efficiency of carbon fixation in chloroplasts. Protein content and, hence, leaf nitrogen status is an important determinant of photosynthesis. Accordingly, leaf nitrogen status would be related to WUE, especially among capacity types (Rao et al., 1995). Specific leaf nitrogen (SLN) expressed as amount of nitrogen present per unit leaf area has emerged as a useful alternative approach for assessing differences in WUE (Sheshshayee et al., 2006). Specific leaf nitrogen (SLN) is computed by multiplying leaf nitrogen (percentage) by SLA. For estimation of leaf nitrogen, refer Section IV of chapter: Nitrogen Compounds and Related Enzymes for detailed description.

6.4 MINERAL ASH CONTENT

Leaf mineral ash is yet another useful parameter often used as an estimate of total transpiration. Several reports verified the relationship between transpiration and mineral ash content. The methodology has been developed well and also used to screen mapping populations, leading to the identification of Quantitative Trait Loci (Merah et al., 1999).

1. Leaf samples are oven dried (70°C for 3 days) and grind in to fine powder using a pestle and mortar or a ball mill. The procedure involves ashing of the samples by complete combustion in a muffle furnace.
2. After recording the empty weight of a silica crucible (W_{EC}), a quantity of finely powdered leaf sample is placed in the crucible and its weight (W_{CL}) determined before placing in a muffle furnace at 600°C for 6–8 h, until the sample is completely oxidized to ash.
3. The crucible is cooled for a sufficient time and weight of the crucible plus ash (W_{CA}) is recorded.
4. The differences between initial and final sample weights indicate the mineral ash content, which can be expressed either in percentage or g kg^{-1}.
5. The ash weight (W_A) and mineral ash content is calculated as follows:

$$W_A = W_{CA} - W_{EC}$$

$$\text{Mineral ash content}\,\% = \left(\frac{W_A}{W_{CL}}\right) \times 100$$

6.5 LEAF ANATOMY

The leaf traits, viz., waxiness, pubescence, rolling, and thickness decrease radiation load to the leaf surface. Benefits include a lower evapotranspiration rate and reduced risk of irreversible photoinhibition. However, they may also be associated with reduced radiation use efficiency, which would reduce yield under more favorable conditions. These traits generally affected high genotype × environment interactions. Hence, reliability and repeatability is low.

6.6 LEAF PUBESCENCE DENSITY

Leaf pubescence is a common feature of xerophytic plants, as well as some crop plants. Generally, leaf pubescence density increases reflectance from the leaf, resulting in lower leaf temperatures under high irradiance. Leaf pubescence density is an important adaptive trait under water-stressed conditions. Densely pubescent lines had increased vegetative vigor, greater root density, and deeper root extension (Garay and Wilhelm, 1983). Increased leaf pubescence density may also increase leaf boundary layer resistance by up to 50%. Reduced leaf temperature, restricted transpiration water loss, and enhanced photosynthesis due to lower radiation penetration into the canopy were also reported to be associated with the dense pubescence trait (Specht et al., 1985).

Five leaves are to be collected from each genotype and cut each leaf into bits of 1 cm^2 with the help of graph paper. Then count the number of trichomes present in each leaf under a stereo microscope and expressed as number of trichomes per 1 cm^2 trichome density.

6.7 DELAYED SENESCENCE OR STAY-GREENNESS

The trait may indicate the presence of drought avoidance mechanisms, but probably does not contribute to yield per se if there is no water left in the soil profile by the end of the cycle to support leaf gas exchange. However, research in sorghum has indicated that stay green is associated with higher leaf chlorophyll content at all stages of development and both were associated with improved yield and transpiration efficiency under drought. They suggest that the rate of senescence rather than onset of senescence is an important component of stay-green.

Leaves senesce early in response to drought and heat stress, particularly when these stresses occur during postflowering stage of seed filling. Some genotypes tolerate drought during grain filling by keeping their leaves green; these cultivars are termed stay-green types. Similarly, the stay-green character and chlorophyll retention in leaves under heat stress conditions was considered expression of heat tolerance (Fokar et al., 1998).

Senescence is scored on a scale from 0 to 10, dividing the percentage of estimated total leaf area that is dead by 10.

1 = 10% dead leaf area
2 = 20% dead leaf area
3 = 30% dead leaf area
4 = 40% dead leaf area
5 = 50% dead leaf area
6 = 60% dead leaf area
7 = 70% dead leaf area
8 = 80% dead leaf area

9 = 90% dead leaf area
10 = 100% dead leaf area

Note: Leaf senescence should be scored on two or three occasions 7–10 days apart during the later part of grain filling.

6.8 LEAF WAXINESS

Excessive deposition of epicuticular wax in crops like sorghum, sugarcane is found to increase leaf reflectance of visible and infrared radiation, decrease net radiation in the field, and decrease cuticular transpiration. Epicuticular wax is therefore an effective component of drought resistance (avoidance mechanism). The protocol given here is hold good for sorghum crop varieties.

Wax content was determined by two methods:

1. Gravimetric analysis
2. Colorimetric analysis

Gravimetric analysis:

The procedure followed that of Silva Fernandez et al. (1964).

1. Four leaf blades are immersed, one at a time, each for 15 s in 100 mL redistilled chloroform.
2. The extract was filtered and evaporated in vacuo at 35°C.
3. The residue was weighed after additional drying for 24 h at room temperature.
4. The amount of wax was calculated against leaf area (both leaf surfaces) of sample, as determined by linear measurements.

Colorimetric analysis:

Principle: This method was developed by Ebercon et al. (1977). The development of the method is based on color change produced due to the reaction of wax with acidic dichromate ($K_2Cr_2O_7$).

Procedure:

1. The acidic dichromate ($K_2Cr_2O_7$) reagent is prepared by mixing 40 mL deionized water with 20 g powdered potassium dichromate.
2. The resulting slurry was mixed vigorously with 1 L concentrated sulfuric acid and heated (below boiling) until a clear solution was obtained.
3. The individual sample consisted of 10 sorghum leaf discs, having a total area (both surfaces) of 30.8 cm^2.
4. Each sample is immersed in 15 mL redistilled chloroform for 15 s.
5. The extract is filtered and evaporated on a boiling water bath, until the smell of chloroform could not be detected.
6. After adding 5 mL of reagent, samples are placed in boiling water bath for 30 min.

7. After cooling, 12 mL of deionized water is added.
8. Several minutes were allowed for color development and cooling and then the optical density of the sample is read at 590 nm.

Standard wax solutions are prepared from carnauba wax (found very similar to sorghum grain wax), carbowax 3000 (Polyethylene glycol-3000), and sorghum wax collected from leaf sheaths of test plants. Waxes are dissolved in redistilled chloroform and 15 mL of aliquots containing a range of concentrations were prepared. These aliquots are carried through the above analytic procedures.

6.9 LEAF ROLLING

Leaf rolling reflects plant moisture deficits and can indirectly reflect presence of soil compaction. As plant moisture content declines, the plant often "protects" itself from excessive plant moisture loss (transpiration) by rolling its leaves. The rolled leaf offers less exposed surface area and transpiration is reduced. Thus, the act of leaf rolling is a sort of defensive posture by the plant against dry weather.

The leaf rolling is scored visually using 1–5 scale as given here:

1 = unrolled, turgid
2 = leaf rim starts to roll
3 = leaf has a the shape of a V
4 = rolled leaf rim covers part of leaf blade
5 = leaf is rolled like an onion leaf

Note: Leaf rolling is to be measured before flowering when leaves are still more upright; leaves are less likely to roll later.

6.10 LEAF THICKNESS (MM)

Leaf thickness was recorded on five randomly selected plants and measured by Digital Vernier Callipers and expressed in millimeter.

6.11 STOMATAL INDEX AND FREQUENCY

The number of stomata per unit area of leaf surface varies considerably among plant species. Leaves of some plants have many stomata and others have few. Environmental conditions of higher light intensity tend to have smaller and more numerous stomata than those in wet and shaded environments. On average, the frequency ranges from less than 2,500–40,000 cm^2 or more.

Stomatal index serves as a reliable index for screening drought-tolerant cultivars.

Procedure:

1. Take a fresh leaf and smear a small quantity of quickfix or nail polish over the leaf evenly as a film.
2. After 15 min, remove the film and place it on a glass slide and observe under the microscope.
3. Count the number of stomata and epidermal cells within the low power field vision. Take three such readings randomly and arrive at an average.
4. Calculate the number of stomata per unit area from the number of stomata observed within the area of field vision. This gives the stomatal frequency.
5. Stomatal index is calculated using the formula

$$\text{Stomatal index}\,(\text{I}) = \frac{\text{Number of stomata}}{\text{Number of epidermal cells}} \times 100$$

6.12 OTHER INDICATORS FOR DROUGHT TOLERANCE

The response of plants to drought may be assumed by drought index based on yield under both stressed and nonstressed conditions (Fischer and Wood, 1981). The drought tolerance index is then calculated as

$$\text{Drought index (DI)} = \frac{\text{Yield under stress}}{\text{Yield under nonstress}}$$

Furthermore, the drought tolerance efficiency (DTE) can be worked out by multiplying DI with 100 as follows:

$$\text{DTE}\,(\%) = \frac{\text{Yield under stress}}{\text{Yield under nonstress}} \times 100$$

6.13 PHENOLOGICAL TRAITS

Flowering time is recognized as the most critical factor to optimize adaptation in environments differing in water availability and distribution during the growing season. Positive associations between plasticity of yield and flowering time across different levels of water availability have been reported in different crops (Sadras et al., 2009).

1. *Flowering date*

   The date when a plot reaches 50% flowering is recorded. To improve the quality of the data, the area to be rated can be restricted to a specific central, fully bordered part of the plot. Estimates of flowering should be recorded at least three times per week. Under drought, lines with later flowering dates will tend to be stressed more than lines that flower early, because the stress intensity increases over time. To correct for this effect, lines can be sown with similar flowering dates in separate experiments and stress applied at the appropriate time for each experiment. Another approach is to make a statistical correction for flowering date. This can be done by using flowering date in the control as a covariate in the analysis.

2. *Flowering delay*
   Flowering delay is best expressed when the stress is severe, so it is easily seen in fields where drying occurs over a period of weeks. It is calculated as follows: Flowering delay = days to flowering in stress treatment – days to flowering in control treatment.
   Note: Because this character is the difference between two independent measurements of flowering date, the error is generally larger for the delay than for the flowering date alone.

CHAPTER

Tissue water related traits

7

Water is the most abundant constituent of most organisms including plants. The actual water content will vary according to tissue and cell type and it is dependent on physiological and environmental conditions. Typically, water accounts for more than 70% by weight. The dominant process in water relations of the whole plant is the absorption of large quantities of water from the soil, its translocation through the plant, and its eventual loss to the surrounding atmosphere as water vapor. Of all the water absorbed by plants, less than 5% is actually retained for growth and even less is used biochemically. Hence, plants need to maintain these dynamic water relations to sustain normal growth and development. Crop plants or genotypes greatly vary in maintaining tissue water content under moisture stress conditions. Adoption level of genotypes which maintain higher tissue water content under these extreme conditions will be high. Hence, in this chapter, various techniques of quantifying water status in various plant parts are described.

7.1 OSMOTIC POTENTIAL

Osmotic adjustment (OA) is defined as the active accumulation of solutes that occurs in plant tissues in response to an increasing water deficit. OA is considered a useful measure because it provides a means for maintaining cell turgor when tissue water potential declines. OA has been shown to maintain stomatal conductance and photosynthesis at lower water potentials, delayed leaf senescence and death, reduced flower abortion, improved root growth, and increased water extraction from the soil as water deficit develops (Turner et al., 2001). Maintenance of cell turgor by osmotic adjustment can decrease the impact of water stress. There is considerable evidence on the role of osmotic adjustment as a mechanism of drought tolerance in several crop species, for example, wheat (Morgan, 2000) and sorghum (Santamaria et al., 1990; Tangpremsri et al., 1995). An increasing number of reports are providing evidence on the association between high rate of osmotic adjustment and sustained yield and biomass under water stress conditions (Blum, 2005).

The water status in soil and leaf are generally recorded to study the effect of drought on physiological process.

Phenotyping Crop Plants for Physiological and Biochemical Traits. http://dx.doi.org/10.1016/B978-0-12-804073-7.00007-7

Osmotic or solute potential is a main component of water potential, which reflects the amount of solutes dissolved in plant tissues. Total concentration of osmotically active dissolved particles in a solution without considering particle size, density, configuration, or electrical charge is known as osmolality. The addition of solute particles to a solvent (water in plants) changes the free energy of the solvent molecules, which allow us indirect means (vapor pressure, freezing point, or boiling point) for the measurement of osmotic potential.

The measurement of osmotic potential can only be made indirectly by comparing one of the solution colligative properties (vapor pressure and freezing point are the most common) with the corresponding cardial property of the pure water. The osmotic potential can be measured by using osmometer based on the depression of the freezing point or by modern osmometer based on the measurement of vapor pressure depression. The vapor pressure decrease in a solution is directly proportional to the amount of solutes added to it (Raoult's law). The measurement of vapor pressure depression is made by thermocouple Hygrometry.

The osmometer has a small chamber to which is sealed a thermocouple hygrometer. A thermocouple measures temperature by the voltage between two dissimilar metals that are joined together. As the vapor pressure equilibrates in the chamber airspace, the thermocouple senses the ambient temperature of the air, thus establishing the reference point of measurement. Under electronic control the thermocouple then seeks the dew point temperature within the enclosed space, giving output proportional to the differential temperature. This difference between the ambient temperature and the dew point temperature is the dew point temperature depression which is an explicit function of solution vapor pressure.

7.1.1 DETERMINATION OF OSMOTIC POTENTIAL USING VAPOR PRESSURE OSMOMETER (PEARCY ET AL., 1989)

7.1.1.1 Procedure

Calibration of the instrument: Initially, the osmometer is calibrated using the standard solution 290, 1000, and 100 mmoles kg^{-1}.

Preparation of calibration curve: Prepare different concentrations of KCl and NaCl solutions and measure the osmolality at various concentrations (Table 7.1).

Table 7.1 Concentration of NaCl and KCl for Preparation of Various Concentrations of Osmolality

Sr. No.	Concentration	NaC1 (MPa)	KC1 (MPa)
1	0.05	−0.232	−0.232
2	0.10	−0.452	−0.452
3	0.20	−0.9901	−0.888
4	0.30	−1.326	−1.326
5	0.40	−1.760	−1.760
6	0.50	−2.190	−2.190

Plot the osmolality values versus osmotic potential values to obtain a calibration curve depicting the relationship between osmolality values and osmotic potential.

Sample preparation

1. Take leaf discs from the leaf samples to be analyzed and cover with aluminum foil and freeze in liquid nitrogen.
2. Thaw the tissue at room temperature for 30 min.
3. Take out the leaf discs and place them in eppendorf tubes, cut the bottom tip of the tube, and place it inside a fresh tube for collecting sap.
4. Spin the samples at 10,000 rpm for 10 min and take the sap collected in the lower tube.
5. Measure the osmotic potential of the sap using vapor pressure osmometer.

Sample loading and measurement

1. Rotate the sample chamber lever and pull out the sample slide.
2. Using forceps place a single sample disc in the central depression of the sample holder.
3. Aspirate the sample using micropipette and with the aid of the notch in the pipette guide place the droplet on the sample disc.
4. Gently push sample slide into the instrument and close the sample chamber lever.

When the measurement is complete chime sound is heard and osmolality (mmoles kg^{-1}) is displayed on the screen. Osmolality in terms of mmoles kg^{-1} can be converted to osmotic potential (MPa).

7.2 LEAF WATER POTENTIAL

Determination of leaf water potential by pressure bomb: Pressure bomb is used for accurate determination of leaf water potential based on the method devised by Scholander et al. (1965). This method is ideally suited for measuring the leaf water potential. It is simple, inexpensive and could be used in the field. Water potential of xylem vessels is normally at a negative tension. If the leaf is placed in a chamber and the cut end is extended out through a seal, pressure can be applied on the surface of the leaf/petiole until the xylem tension is balanced and water is forced back to the cut surface. Since leaves continue to transpire after detachment from the plant leading to loss of leaf water, the leaf should be inserted in the pressure chamber as quickly as possible.

Principle: The pressure is applied to a detached leaf to return the water interface, where it was before detachment, is equal and opposite to the tension in the xylem of the intact plant. Because the osmotic potential of the xylem sap is usually less than 0.02 MPa, the hydrostatic pressure in the xylem is equal to the water potential.

Procedure:

- Leaves, after excision at petiole, are put into butter paper bags. All such bags are enclosed in a polythene envelope.
- The envelope, after wrapping carefully, is kept in a thermocol ice box containing ice cubes. This precaution is essential to prevent any further decline in leaf water potential due to desiccation.
- A sample leaf is inserted in a Pressure Chamber apparatus with petiole protruding out from the air-tight gasket.
- Compressed, dry nitrogen gas is passed into the chamber slowly but constantly through a flow regulator until the xylem sap oozes out at the cut end of the petiole.
- Carefully increase the pressure in the chamber while observing the out end of the leaf from the side with a magnifying hand lens.
- Increase the pressure one bar every 4 or 5 s.
- When xylem sap first appears through the cut surface, cut-off the air inlet valve and read the gauge, indicate the pressure required with a negative sign. It is an estimate of xylem water potential.
- Take enough readings till a constant measurement is obtained.
- This instrument measures water potential in bars. However, bar may be converted in Pascals as follows:

$$1 \text{ bar} - 10^5 \text{ Pascal} = 0.1 \text{ Megapascal.}$$

7.3 RELATIVE WATER CONTENT

Refer chapter: Other Drought-Tolerant Traits for detailed description.

7.4 CELL MEMBRANE INJURY

7.4.1 CELL MEMBRANE PERMEABILITY BASED ON LEAKAGE OF SOLUTES FROM LEAF SAMPLES

Plant responses to drought at cellular level depend upon (1) the degree of water deficit, (2) the duration of stress period, and (3) the genotypes. The cell membrane is the primary site of damage and it results in an increased leakage of solutes/ions through the membrane under water stress condition. A genotype that can maintain membrane permeability would also exhibit higher levels of intrinsic stress tolerance. Hence, ion leakage may be used as an index for screening plant species against drought to heat stress.

Principle: Membrane permeability is estimated by measuring the extent of ion leakage from cells through either determining absorbance at 273 nm or measuring electrical conductivity using a conductivity bridge. More electrical conductivity means more leakage of solutes/ions and thereby more injury to the membrane. The percent conductivity is equivalent to membrane injury index.

Procedure: A known weight of the leaf sample (0.1 g) cut into pieces of uniform size is taken in test tubes and is incubated in 10 mL of water for 3 h, and the leakage is recorded by reading the initial absorbance at 273 nm, using a spectrophotometer. The samples are then incubated in a hot water bath (100°C) for 15 min and the final absorbance at 273 nm is recorded (Towill and Mazur, 1975). The percent membrane injury is calculated as follows:

$$\%\,\text{Conductivity} = (\text{initial absorbance} \,/\, \text{final absorbance}) \times 100$$

CHAPTER

Heat stress tolerance traits

8

High temperature is one of the major abiotic stresses that adversely affect crop growth at different stages of development. Some researchers believe that night temperatures are major limiting factors; others have reported that day and night temperatures do not affect the plant independently and that diurnal mean temperature is a better predictor of plant response to high temperature with day temperature having a secondary role (Peet and Willits, 1998).

Improvement of high-temperature tolerance is considered vital for yield improvement in many regions and cropping systems. Hence, it is important to identify a reliable protocol under controlled field conditions that allows simultaneous screening of multiple genotypes. On the basis of the temperature stress response, several techniques have been developed to assess the stress effects. This chapter deals with different reliable techniques of phenotyping crop plants/genotypes for thermotolerance.

8.1 CANOPY TEMPERATURE

Canopy temperature is measured remotely by an infrared thermometer (IRT) which is an inexpensive device. Canopies emit long-wave infrared radiation as a function of their temperature. The IRT senses this radiation and converts it into an electrical signal, which is displayed as temperature. Using the thermometer properly is crucial to obtaining reliable data. The most important point in the protocol for using an IRT in breeding nurseries is explained later.

The correlation between canopy temperature and plant water status becomes stronger as plant water status is reduced. Therefore, measurements should be made under well-developed drought stress—typically when most of the material in the nursery presents some leaf wilting or leaf rolling at midday.

Measurements should be done at or just after the solar noon, when the plant water deficit is maximized. Since the plant water status changes over the day, measurements on large populations must be done within about 2 h. The thermometer has a fixed angle of view (ca 2–5 degrees, depending on the model). Therefore, the size of the measured target area depends on the distance between the thermometer and the target. Distance, position, and angle of measurement with respect to viewed plot must be maintained with all plots measured.

The target must consist only of canopy leaves. Any other object in the target area, such as soil surface or panicles, will result in a temperature reading that does not represent the leaf canopy temperature. Soil is generally hot and cereal inflorescences

Phenotyping Crop Plants for Physiological and Biochemical Traits. http://dx.doi.org/10.1016/B978-0-12-804073-7.00008-9

(panicles or spikes) are much warmer than leaves because they transpire very little. For this reason, screening canopy temperature measurements under drought stress can be done only after full ground cover has been attained and prior to inflorescence emergence. Since the assessment of plant stress by canopy temperature within a breeding population is relative, atmospheric conditions during measurements should be relatively stable. Cloudy or windy conditions should be avoided. Transient cloudiness is especially difficult since it has an immediate effect on leaf temperature. Viewing solar spectral reflectance from the canopy will not harm the instrument but may bias temperature measurement. Therefore, readings should be made with sun behind the operator—basically similar to the rule for photography. This should be taken into account when the nursery layout is planned. The nursery should contain a running check (control) cultivar, every 10–100 genotypes, depending on the case. The canopy temperature of the running check provides a basis for assessing site variability and offers a means for normalizing data against such variability. Experience shows that if work is performed carefully as outlined earlier, about 1.5–2.0°C can be the least significant difference (at 5%). If stress is sufficient and atmospheric demand for transpiration is high, genotypes may differ by up to 5–10°C on any given day, depending on the crop and the nature of the population.

Measurements should be performed several times during the drying cycle, once or twice a week, depending on the progress of stress. For each date of measurement, data can be processed in three forms: actual temperature, temperature of the genotypes as a percentage of the mean temperature of the block, and temperature of the genotypes as a percentage of the temperature of the nearest running check. The final data by which selection is performed are usually derived from the day with the largest variation among genotypes, which is the date of maximum plant water deficit at peak stress.

8.2 CHLOROPHYLL STABILITY INDEX (CSI)

Refer chapter: Other Drought-Tolerant Traits for detailed description.

8.3 CHLOROPHYLL FLUORESCENCE

The procedure used to measure chlorophyll fluorescence characteristics is similar to that of Smillie and Hetherington (1990) using Fluorimeter or Photosynthetic efficiency analyzer system. Chlorophyll fluorescence measured by photosynthetic efficiency analyzer in groundnut (Babitha et al., 2006; Pranusha et al., 2012), in rice (Renuka Devi et al., 2013), and in Blackgram (Sudhakar et al., 2006) in this laboratory.

1. For each main unit and replicate, four leaves of the third fully expanded leaf from the top of the main axis are detached and evenly distributed among four cap tubes containing 1 mL of distilled water.

2. Four more leaves from each of the additional plants are detached and distributed among the four tubes.
3. One tube is designated for each of the three different high-temperature treatments (45, 50, and 55°C).
4. The tube that is not exposed to heat is treated as control. The tubes were capped and placed in water baths maintained at selected temperatures for 5 min.
5. After the high-temperature treatment, leaves are dark adapted for 30 min at room temperature.
6. Chlorophyll fluorescence is recorded with fluorescence measurement system (Handy Photosynthetic Efficiency Analyzer (PEA), Hansatech Electronics Ltd., UK).
7. The equipment displays initial (F_0) and maximum (F_m) fluorescence values. Variable fluorescence (F_v) is derived by subtracting F_0 from F_m. F_v/F_m value is then calculated for analyzing Photosystem II efficiency of a plant genotype.

8.4 THERMO INDUCTION RESPONSE (TIR) TECHNIQUE

The magnitude of heat stress effect on plants is variable at field screening conditions and the results can be inconsistent and seasonally limited. Hence, it is important to develop a reliable protocol under controlled conditions that allows simultaneous screening of multiple genotypes.

To cope with changing environmental conditions, plants synthesize a set of stress-responsive proteins that would be necessary for altering specific metabolic processes leading to adaptation to the stress. These adaptive responses are primarily associated with the expression of specific heat shock proteins (HSPs; Chen and Asada, 1990; Cushman and Bohnert, 2000).

The expression of these HSPs is dependent on finely regulated activation by heat shock transcription factors (HSFs; Scharf et al., 1998). Genetic variability in stress adaptation is dependent on the differential expression of these stress-responsive genes (Fender and O'Connell, 1990; Krishnan et al., 1989). There is convincing evidence for the fact that these stress-responsive genes are predominantly expressed when plants experience sublethal levels of stress. This would bring in certain specific alterations in metabolism, leading to increased tolerance when plants subsequently experience a lethal level of the stress.

Genetic variability in stress tolerance is, therefore, a result of the extent of stress gene expression when plants experience such lethal stresses. This phenomenon has been extensively studied for high-temperature stress tolerance. Based on the knowledge that a gradually increasing induction stress would trigger the expression of specific genes leading to a greater level of tolerance to severe stress (Sun et al., 2001), a novel temperature induction response (TIR) technique has been developed and standardized to assess the genetic variability in acquired thermo tolerance as a good indication of intrinsic stress tolerance at the cellular level. Any genotype that shows

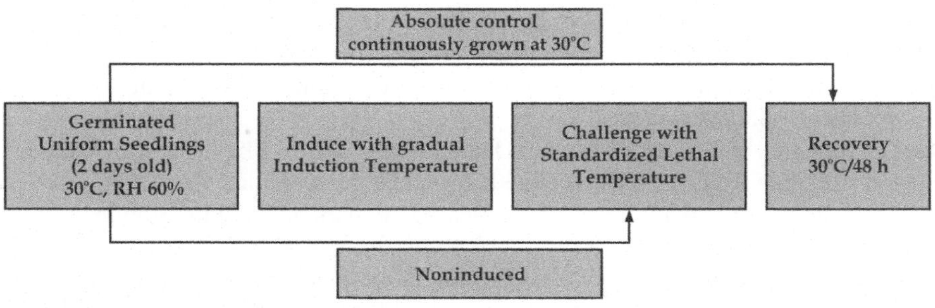

FIGURE 8.1 General Protocol for TIR.

superior expression of stress responsive genes would acquire higher tolerance to more severe stress levels and hence can be considered as having higher intrinsic tolerance at the cellular level. This technique capitalizes on acquired stress tolerance at a sublethal induction temperature.

Procedure:
This approach of TIR involves first the identification of challenging temperature and induction temperature and later standardizing them before being used for screening material for intrinsic tolerance (Fig. 8.1).

1. *Identification of lethal temperature treatment:* To assess the challenging temperatures for 100% mortality, 48-h-old seedlings of any crop are exposed to different lethal temperatures (48–60°C) for varying durations (1–4 h) without prior induction. Thus exposed seedlings were allowed to recover at 30°C and 60% relative humidity for 48 h at the end of recovery period. Record percent mortality of genotypes after recovery. The temperature at which maximum mortality of seedlings has recorded is considered lethal temperature to that particular crop.
2. *Identification of sublethal (induction) temperatures:* During the induction treatment, expose the seedlings to a gradual increase in temperature for a specific period. These temperature regimes and duration that are varied from crop to crop are to be standardized. The temperature regimes and durations are varied to arrive at optimum induction protocol. The optimum sublethal temperatures are arrived based on the percent survival of seedlings. The sublethal treatment that recovered least percent seedlings survival reduction is considered optimum temperature (Sudhakar et al., 2012; Renuka Devi et al., 2013).
3. *Thermo Induction Response (TIR):*
   a. Surface sterilize seeds by treating with systemic fungicide solution for 30 min and wash with distilled water for four to five times.
   b. Then keep seeds for germination at 30°C and 60% relative humidity in the incubator.

c. After 42 h (depends on crop seed) select uniform seedlings in each genotype and sow in aluminum trays (similar gauge) filled with soil.
d. These trays with seedlings are to be subjected to sublethal temperatures and gradual temperature (prestandardized) in the Programmable Plant Growth Chamber.
e. Thus exposed seedlings (trays) are exposed to lethal temperatures or induced prestandardized period.
f. Another subset of seedling trays are directly exposed to lethal temperatures (noninduced).
g. Induced and noninduced ragi seedlings are allowed to recover at 30°C and 60% relative humidity for 48 h.
h. Maintain a control tray at 30°C, without exposing to sublethal and lethal temperatures.

The following parameters were recorded from the seedlings

1. % survival of seedlings =

$$\frac{\text{No. of seedlings survived at the end of recovery}}{\text{Total no. of seedlings sown in the tray}}$$

2. % reduction in root growth =

Actual root growth of control seedlings – (Actual root growth of treated seedlings/ Actual root growth of control seedlings) × 100

3. % reduction in shoot growth =

Actual shoot growth of control seedlings – (Actual shoot growth of treated seedlings/ Actual shoot growth of control seedlings) × 100

8.5 MEMBRANE STABILITY INDEX

8.5.1 MEMBRANE PERMEABILITY BASED ON LEAKAGE OF SOLUTES FROM LEAF SAMPLES

Abiotic stresses can cause a significant alteration in membrane composition and hence its permeability characteristics. Therefore, a genotype that can maintain membrane permeability would also exhibit higher levels of intrinsic stress tolerance. Membrane permeability is estimated by measuring the extent of ion leakage from cells through either determining absorbance at 273 nm or measuring electrical conductivity using a conductivity bridge. A known weight of the leaf sample is incubated in water for 3 h, and the leakage is recorded by reading the initial absorbance at 273 nm, using a spectrophotometer. The samples are then incubated in a hot water bath (100°C) for 15 min and final absorbance is recorded at 273 nm (Towill and Mazur, 1975). The percentage leakage is calculated as follows:

% leakage = (initial absorbance/final absorbance) × 100.

CHAPTER

Oxidative stress tolerance traits

9

9.1 OXIDATIVE DAMAGE

Light is the most crucial input for primary photosynthetic processes. However, when the absorbed radiant energy exceeds the plant's capacity to utilize it for photosynthesis, an energy imbalance occurs in the chloroplast thylakoid and normally causes damage to photo system II (PS II). The capacity to utilize light energy normally decreases when plants are under stress. Reduction in photosynthetic capacity is normally associated with: (1) a decrease in stomatal conductance resulting in reduced substrate CO_2 influx; and (2) a stress-induced reduction in the chloroplast efficiency in fixing carbon. These, in turn, result in a lack of utilization of photochemical energy leading to an enhanced NADPH/NADP+ (reduced to oxidized nicotinamide adenine dinucleotide phosphate) ratio. The excessive excitation energy is then used to reduce O_2 and produce reactive oxygen species (ROS).

A series of reactions involving electron transfer or excitation energy transfer are triggered under oxidative stress conditions leading to the production of several ROS. The ROS are produced either through the monovalent reduction of molecular oxygen to produce the super oxide anion (O_2^-) or through energy transfer from a triplet excited chlorophyll molecule to ground-state oxygen to form singlet oxygen ($^1O_2$) (Asada, 1994; Bowler et al., 1992). Furthermore, the dismutation of the superoxide catalyzed by superoxide dismutase (SOD) generates hydrogen peroxide radicals (H_2O_2). If these moieties are not scavenged, their interaction with metal ions will result in the production of the most potentially reactive species of oxygen, the hydroxyl radical (OH^-).

Production of these highly reactive and potentially damaging species of oxygen leads to a series of degenerative processes, collectively called "oxidative stress damage" (Asada, 1994; Foyer et al., 1994; Inze and Van Montague, 1995). If the events leading to the excessive generation of damaging ROS go unchecked, senescence sets in as an inevitable consequence, severely reducing productivity. Thus, it is imperative that plants maintain a balance between the energy absorbed through the primary photochemistry and its subsequent utilization in metabolism (Udayakumar et al., 1999).

To adapt to these conditions, plants have evolved several mechanisms, such as avoidance of excess light interception, dissipation of excess excitation energy, and management of excess photochemical energy. Crop species and genotypes exhibit significant variation in oxidative stress tolerance, owing to the variation in the

Phenotyping Crop Plants for Physiological and Biochemical Traits. http://dx.doi.org/10.1016/B978-0-12-804073-7.00009-0

scavenging strategy adopted, as well as the difference in the efficiency of such strategies. Hence in this chapter reliable protocols are given to quantify various oxidative enzymes for phenotyping crop plants for oxidation stress tolerance

9.1.1 ANTIOXIDANT ENZYMES

Drought stress affects many physiological processes of plants leading to accumulation of reactive oxygen species (ROS) such as superoxide radical, hydrogen peroxide, and hydroxyl radical. These ROS inhibit protein synthesis by oxidation of mRNA. Therefore, plants must adapt certain mechanisms to scavenge these ROS. This oxidative damage in the plant tissue is alleviated by a concerted action of both enzymatic and nonenzymatic antioxidant metabolisms. These mechanisms include β carotenes, ascorbic acid, reduced glutathione, and enzymes including superoxide dismutase, peroxidase, catalase, etc. There are many reports that enhanced antioxidant enzyme activity increased resistance to environmental stresses. Hence, estimation of antioxidant enzyme activities is essential to know the oxidative stress tolerance in plants.

9.2 SUPEROXIDE DISMUTASE (SOD)

The cell generates a variety of molecules during its metabolic processes. Environmental stresses such as high/low temperature, water stress, air pollution, ultraviolet light, and chemicals result in the excess production of active species such as super oxide, hydrogen peroxide, and hydroxyl radicals. Unless these toxic molecules are eliminated, damage to the macromolecules such as DNA/tissue is imminent. SOD is conveniently assayed using a slightly modified procedure (Madamanchi et al., 1994) originally described by Beauchamp and Fridovich (1971).

Principle: Superoxide dismutase (SOD), a metal-containing enzyme, plays a vital role in scavenging superoxide (O_2^-) radical.

$$O_2^- + 2H^+ \rightarrow H_2O_2 + O_2$$

Hydrogen peroxide is eliminated by peroxidases and catalases. Superoxide dismutase activity was determined by measuring its ability to inhibit the photochemical reduction of nitro blue tetrazolium (NBT). The reaction mixture lacking enzyme develop maximum color and color intensity decreased with increase in the enzyme activity.

Chemicals required:

- Potassium monohydrogen phosphate (K_2HPO_4)
- Potassium dihydrogen phosphate (KH_2PO_4)
- Methionine
- Riboflavin
- EDTA
- Nitro Blue Tetrazolium (NBT)

Preparation of reagents:

- *Potassium phosphate buffer stock:*
  Solution A: potassium monohydrogen phosphate (250 mM): 4.37 g in 100 mL.
  Solution B: potassium dihydrogen phosphate (250 mM): 3.402 g in 100 mL.
  Add solution A to solution B with constant stirring and pH 7.8 is maintained using sodium hydroxide
- Potassium phosphate buffer, 50 mM with pH: 7.8 (100 mL of stock in 400 mL of distilled water)
- Methionine, 100 mM: 298 mg in 20 mL of D.D.H_2O
- Riboflavin, 10 mM: 37.6 mg in 10 mL of D.D.H_2O
- EDTA, 5 mM: 93 mg in 100 mL of D.D.H_2O
- NBT, 750 mM: 6.1 mg in 10 mL of D.D.H_2O

Extraction:

1. Fresh sample of 1 g is grinded with 10 mL of 50 mM potassium phosphate buffer (pH:7.8) in precold mortar using pestle at 4°C.
2. Then the sample is centrifuged at 10,000 rpm for 10 min.
3. Collect the supernatant. Store in deep freezer.

Estimation:

Preparation of 3 mL cocktail solution:

1. 0.6 mL of 250 mM potassium phosphate buffer, 0.39 mL of 100 mM Methionine, 0.0006 mL of 10 mM riboflavin, 0.06 mL of 5 mM EDTA, 0.3 mL of 750 mM NBT, and 50 μL of enzyme extract were taken into a test tube and make up to 3 mL with distilled water.
2. Prepare cocktail solution freshly and keep under fluorescent bulb for 15 min.
3. Then read absorbance (OD) at 560 nm by UV–VIS spectrophotometer using kinetics method.
4. Preparation without enzyme extract and NBT serve as a blank to calibrate the spectrophotometer. Set another control having NBT but no enzyme extract as reference control.
5. Calculate the % inhibition.
6. The 50% inhibition of the reaction between riboflavin and NBT in the presence of methionine is taken as one unit of SOD activity.
7. The enzyme activity is expressed as OD min^{-1} g^{-1}.

   Calculation: (Maximum absorbance – Minimum absorbance) × 60 × 2.

9.3 CATALASE

Catalase is an enzyme present in nearly all plant and animal cells. Catalase has a double function as it catalyses the following reactions:

1. Decomposition of hydrogen peroxide to give water and molecular oxygen

$$H_2O_2 \rightarrow 2H_2O + O_2$$

2. Peroxidation of H donors (methanol, formic acid, phenol) with consumption of one mole of peroxide.

$$ROOH + AH_2 \rightarrow H_2O + ROH + A\,(\text{peroxidative})$$

Principle: The UV light absorption of hydrogen peroxide solution can be easily measured between 230 and 250 nm. On decomposition of hydrogen peroxide by catalase, the absorption decreases with time. The enzyme activity could be arrived at from this decrease. But this method is applicable only with enzyme solution which do not absorb strongly at 230–250 nm (Luck, 1974).

Chemicals required:

- Monobasic sodium phosphate (NaH_2PO_4)
- Dibasic sodium phosphate (Na_2HPO_4)
- Polyvinyl pyrrolydine (PVP)
- Hydrogen peroxide (H_2O_2)

Preparation of reagents:

Preparation of 0.05 M phosphate buffer with pH: 7.2

Solution A: Monobasic sodium phosphate (NaH_2PO_4): 0.6 g in 100 mL of distilled water.
Solution B: Dibasic sodium phosphate (Na_2HPO_4): 0.7 g in 100 mL of distilled water.

Add solution A of 28.0–72.0 mL of solution B and make up to a total volume of 200 mL with distilled water.

50 mM phosphate buffer with pH 7.0

Solution A: Monobasic sodium phosphate (NaH_2Po_4): 0.6 g in 100 mL of distilled water.
Solution B: Dibasic sodium phosphate (Na_2HPo_4): 0.7 g in 100 mL of distilled water.

Add solution A of 39.0–61.0 mL of solution B and dilute to a total volume of 200 mL with distilled water:

- 1% polyvinyl pyrrolydine (PVP): 1 g in 100 mL of distilled water
- 0.03% hydrogen peroxide: 0.03 mL in 100 mL of distilled water

Extraction:

1. Fresh sample of 300 mg fresh sample is grinded with 2.5 mL of 0.05M sodium phosphate buffer (pH: 7.0) and 1 mL of 1% PVP in precold mortar using pestle at 4°C.
2. Then the sample is centrifuged at 10,000 rpm for 15 min at 4°C.
3. Collect the supernatant.

Estimation:

1. 50 mM buffer solution of 2 mL, 0.95 mL of 0.03% hydrogen peroxide, and 0.05 mL of enzyme extract were taken into a test tube and the resultant is mixed well.
2. Then read absorbance (OD) at 240 nm by UV–VIS spectrophotometer using kinetics method.

3. Preparation without enzyme extract serves as a blank to calibrate the spectrophotometer.
4. The enzyme activity is expressed as units min^{-1} g^{-1}.

   Calculation: (Maximum absorbance – Minimum absorbance) × 60 × 2.

9.4 PEROXIDASE (POD)

Peroxidase (POD) includes in its widest sense a group of specific enzymes such as NAD-peroxidase, NADP-peroxidase, fatty acid peroxidise, etc., as well as a group of very nonspecific enzymes from different sources which are simply known as POD (donor: H_2O_2-oxidoreductase). POD catalyzes the dehydrogenation of a large number of organic compounds such as phenols, aromatic amines, hydroquinones, etc. POD occurs in animals, higher plants, and other organisms.

Principle: Guaiacol is used as substrate for the assay of peroxidase.

$$\text{Guaiacol} + H_2O_2 \rightarrow \text{oxidized guaiacol} + 3H_2O$$

The resulting oxidized (dehydrogenated) guaiacol is probably more than one compound and depends on the reaction conditions. The rate of formation of guaiacol dehydrogenation product is a measure of the POD activity and can be assayed spectrophotometrically at 436 nm (Putter, 1974; Malik et al., 1980).

Chemicals required:

- Monobasic sodium phosphate (NaH_2PO_4)
- Dibasic sodium phosphate (Na_2HPO_4)
- Guaiacol solution
- Hydrogen peroxide solution
- Sodium hydroxide

Preparation of reagents:

Phosphate buffer, 0.1M (pH: 7.0):

Solution A: Monobasic sodium phosphate (NaH_2PO_4)
Solution B: Dibasic sodium phosphate (Na_2HPO_4)
Add solution A of 39 mL to solution B of 61 mL and make up to a total volume of 200 mL with distilled water

- 20 mM Guaiacol solution.
- Hydrogen peroxide solution (0.042% = 12.3 mM): dilute 0.14 mL of H_2O_2 to 100 mL of distilled water.

Extraction:

1. Fresh sample of 1 g is grinded with 3 mL of 0.1 M sodium phosphate buffer (pH: 7.0) in precold mortar using pestle at 4°C.
2. Then the sample is centrifuged at 18,000 rpm for 15 min at 5°C.
3. Collect the supernatant.

Estimation:

1. Buffer solution of 3 mL, 0.05 mL of guaiacol, 0.03 mL of hydrogen peroxide, and 0.1 mL of enzyme extract were taken into a test tube and the resultant is mixed well.
2. Then read absorbance (OD) at 436 nm by UV–VIS spectrophotometer using kinetics method.
3. Preparation of without enzyme extract serves as a blank to calibrate the spectrophotometer.
4. The enzyme activity is expressed as units min^{-1} g^{-1}.

   Calculation: (Maximum absorbance – Minimum absorbance) × 60 × 2.

9.5 FREE RADICALS

1. *Superoxide (O_2^-) ion*

   In addition to the above-mentioned antioxidative enzymes, we can also measure the rate of production of H_2O_2 and superoxide anions (O_2^-) in plants. For estimation of superoxide anions, a simple spectro-photometric method as described by Chaithanya and Naithani (1994) can be used. This method is based on the measurement of superoxide ions by their capacity to reduce nitro-blue tetrazolium solution.

   Chemicals required:

   - Diethyl dithiocarbamate
   - Nitro-Blue Tetrazolium solution (NBT)
   - Monobasic sodium phosphate (NaH_2PO_4)
   - Dibasic sodium phosphate (Na_2HPO_4)

   Preparation of reagents:

   - 1 mM Diethyl dithiocarbamate
   - Nitro-blue tetrazolium solution (0.25 mM): 3.73 mg in 100 mL of D.D.H_2O
   - Sodium phosphate buffer (100 mM; pH 7.2)
   - Solution A: monobasic sodium phosphate (NaH_2PO_4)
   - Solution B: dibasic sodium phosphate (Na_2HPO_4)
   - Add solution A of 39 mL to solution B of 61 mL, adjust the pH 7.2 and then make up to total volume of 200 mL with D.DH_2O.

   Extraction:

   1. Fresh sample of 0.5 g is homogenized under N_2 atmosphere at 0–4°C in 10 mL of sodium phosphate buffer (pH 7.2) containing 1 mM diethyl dithiocarbamate to inhibit superoxide dismutase activity.
   2. After centrifugation at 20,000 × g for 20 min, the supernatant is stored for estimation of superoxide anions.

Estimation:

1. The assay mixture, in a total volume of 3 mL, contains 2.85 mL of sodium phosphate buffer (100 mM; pH 7.2 with 1 mM diethyl dithiocarbamate), 100 μL of NBT (0.25 μM), and 50 μL of supernatant.
2. The absorbance of the end product is measured at 540 nm wavelength using a spectrophotometer.
3. Formation of superoxide anions is expressed as ΔA540 min^{-1} mg^{-1} protein.
4. All the experiments should be carried out in sealed tubes under N_2 atmosphere to minimize oxidation and generation of reactive oxygen species (ROS).

2. *Hydrogen peroxide (H_2O_2)*

Hydrogen peroxide (H_2O_2) content in plant samples can be estimated spectrophotometrically following the method described by Mukherjee and Choudhari (1983) using titanium reagent. The standard H_2O_2 is used for calibration and H_2O_2 content is expressed as μmol H_2O_2 g^{-1} fr.wt.

Chemicals required:

- Titanium oxide (TiO_2)
- Di potassium sulfate (K_2SO_4)
- Sulfuric acid (H_2SO_4)
- Liquid ammonia (NH_3)
- Acetone (CH_3COCH_3)

Preparation of reagents:

Titanium reagent: Titanium oxide of 1.0 g and 10 g K_2SO_4 are mixed and digested with 150 mL of concentrated H_2SO_4 for 2–3 h on a hot plate. The digested mixture is cooled and diluted to 1.5 L with distilled water.

Sulfuric acid (1.0 M)

Extraction:

1. Plant sample (0.5 g) is homogenized in 10 mL of cold acetone.
2. The homogenate is filtered through what man No. 1 filter paper.
3. Titanium reagent (4 mL) is added to whole extract followed by 5 mL of ammonia solution to precipitate the hydrogen peroxide–titanium complex.
4. It is centrifuged for 5 min at 10,000 × g.
5. The supernatant is discarded and the precipitate is dissolved in 10 mL of 1.0 M sulfuric acid.
6. It is recentrifuged to remove undissolved material.

Estimation: Absorbance is noted at 415 nm against blank. Concentration of H_2O_2 is determined using standard curve plotted with known concentration of H_2O_2.

CHAPTER

Salinity tolerance traits

10

Salinity is a major problem in arid and semiarid tropics. In India about 8.6 mha (Pathak, 2000) of land area is affected by soil salinity. Thus, there is a need to develop high yielding varieties or hybrids having salt-tolerant traits/mechanisms. The leaf proline content (LPC) and chlorophyll stability index (CSI) are the two important physiological traits that are directly related to salt stress. The high accumulations of proline and chlorophyll in plants indicate that the plant did not have much problem to survive in salt stress environment. The high LPC and CSI showed that the plant has the ability to convert glutamate into proline and this proline seems to have diverse roles under different abiotic stresses (Verma, 1999). The high CSI value indicates the better availability of chlorophyll in the plant and helps the plant to withstand stress. This leads to increased photosynthetic rate, more dry matter production, and higher productivity. The LPC and CSI served as potent characters in identifying tolerant genotypes to salt stress situation. Apart from these enzymes, phyto antioxidative enzymes play a role in adoption of tolerance to salinity conditions. Protocols for quantification of these enzymes in crop plants are given in this chapter.

10.1 CHLOROPHYLL STABILITY INDEX

Refer to chapter: Other Drought-Tolerant Traits for detailed description.

10.2 PROLINE

Proline is a basic amino acid found in high percentage in basic proteins. Free proline is said to play a role in plants under stress conditions. Though the molecular mechanism has not yet been established for the increased level of proline, one of the hypotheses refers to breakdown of proteins into amino acids and conversion to proline for storage. Many workers have reported several-fold increase in the proline content under physiological and pathological stress conditions. Hence, the analysis of proline in plants has become routine in pathology and physiology divisions of agricultural sciences.

Principle: Free proline from plant tissues may be selectively extracted in aqueous sulphosalicylic acid and its amount is estimated by ninhydrin method. During selective extraction with aqueous sulphosalicylic acid, proteins are precipitated as a

Phenotyping Crop Plants for Physiological and Biochemical Traits. http://dx.doi.org/10.1016/B978-0-12-804073-7.00010-7

complex. The extracted proline is made to react with ninhydrin under acidic conditions to form a red color which is measured at 520 nm (Bates et al., 1973; Chinard, 1952).
Chemicals required:

- Acid ninhydrin
- Aqueous sulphosalicylic acid (3%)
- Glacial acetic acid
- Toluene
- Proline

Preparation of reagent:
Acid ninhydrin:

Dissolve 1.25 g ninhydrin in a mixture of 30 mL warm glacial acetic acid and 20 mL of 6 M Ortho phosphoric acid (OPA), with agitation until it is dissolved. Stored at 4°C and use within 24 h (6 M OPA: 7.64 mL of OPA made up to 20 mL with H_2O).
Extraction:

1. Extract 0.5 g of plant material by homogenizing in 10 mL of 3% aqueous sulphosalicylic acid.
2. Centrifuge the homogenate at 6000 rpm for 30 min and collect the supernatant.

Estimation:

1. Take 2 mL of filtrate in a test tube and add 2 mL of glacial acetic acid and 2 mL of acid – ninhydrin.
2. Heat it in boiling water bath for 1 h.
3. Terminate the reaction by placing the tube in ice bath.
4. Add 6 mL toluene to the reaction mixture and stir well for 20–30 s.
5. Separate the toluene layer and warm to room temperature. Measure the red color intensity at 520 nm.
6. Dissolve 50 mg proline in 50 mL of distilled water in a volumetric flask.
7. Take 10 mL of this stock standard and dilute to 100 mL in another volumetric flask for working standard solution.
8. A series of volumes from 0.1 to 1 mL of this standard solution gives a concentration range 10–100 μg.
9. Then proceed as that of the sample and read the color.

Calculation: Draw a standard curve using absorbance versus concentration. Find out the concentration of proline in the sample using standard regression equation and express as microgram per gram fresh.

10.3 SODIUM (Na) AND POTASSIUM (K) RATIO

10.3.1 POTASSIUM (K)

Principle: The most commonly used method for K determination is by flame photometry. It is based on principle that the atoms of specific element take energy from flame and get excited to the higher orbit. Such atoms release energy of a wavelength that is specific for that element and is proportional to the concentration of atoms of that element.

Instrumentation: The element to be determined is introduced into flame photometer in a solution form. A fine aerosol is formed and the atoms get excited by taking energy from flame created by mixture of liquid petroleum gas mixed with air. The emitted radiation may be of several lengths. It is therefore passed through filters to isolate radiation of desired wavelength. The isolation of radiation can also be achieved by a prism or monochromater. Radiation is measured either by photocell or photomultiplier tube. The concentration of K is measured by comparing the radiation emitted by a known standard with that of a sample. Most flame photometers manufactured give a linear range up to 5 ppm and therefore a desired solution has to be carried out before the determination of K in plant samples.

Standard stock solution: To prepare a stock solution, 1.9069 g of analytical grade KCl is dissolved in deionized water and volume made up to 1 L. This solution contains 1000 ppm K. Prepare 100 ppm K solution by diluting the 1000 ppm K solution 10 times (10 mL in 100 mL final volume). Final standard solutions of 0, 5, and 10 ppm are prepared from 100 ppm K.

Standard curve: To prepare standard curve, the instrument is set at highest concentration of 5 ppm using a standard filter. The manufacturers specify the linear range of K (normally 5 ppm) and suitable factor is calculated for finding out K in plant samples.

Estimation of K in plant samples: The plant samples for K estimation can be digested by diacid.

Di-acid digestion (Tandon, 1993): It is carried out using 9:4 mixture of HNO_3:$HClO_4$. Plant material of 1 g is powdered and placed in 100 mL volumetric flask. To this, 10 mL of acid mixture is added and the content of the flask is mixed by swirling. The flask is placed on low heat hot plate in a digestion chamber. Then, the flask is heated at higher temperature until the production of red NO_2 fumes ceases. The contents are further evaporated until the volume is reduced to about 3–5 mL but not to dryness. The completion of digestion is confirmed when the liquid becomes colorless. After cooling the flask, add 20 mL of deionized or glass distilled water. Volume is made up with deionized water and the solution is filtered through Whatman No. 1 filter paper. The digest is diluted to the suitable concentration range so that final concentration lies between 0 and 5 ppm. The samples are then read in flame photometer at 548 nm wavelength or using filter for K.

10.3.2 SODIUM (Na)

Principle: The procedure for Na is similar to that of potassium. However, a different filter meant for Na has to be used as the radiation emitted by excitation of Na atoms is of different wavelength.

Preparation of standard: Dissolve 2.541 g of NaCl in 1000 mL deionized or distilled water to get 1000 ppm Na. Ten milliliter of this stock solution is diluted to 100 mL to get 100 ppm Na. From this 100 ppm stock solution, a final standard solution of 10 ppm is prepared. If Na is determined by flame photometer fitted with monochromater, the element can be determined at 598 nm wavelength.

Estimation of Na in plant samples: The concentration of Na in plant sample is determined through diacid digestion of samples. The digest is diluted to suitable concentration so that final concentration lies between 0 and 10 ppm. The samples are then fed in a flame photometer and Na concentration read from the standard curve.

10.4 ANTIOXIDATIVE ENZYMES

Refer to chapter: Oxidative Stress Tolerance Traits for detailed description of Super Oxide Dismutase, Catalase, and Peroxidase enzyme activities.

SECTION IV

CHAPTER

Kernel quality traits

11

Quality refers to the suitability or fitness of an economic plant in relation to its end use. Quality varies according to our needs from the viewpoint of seeds, crop growth, crop product, postharvest technology, consumer preferences, cooking quality, keeping quality, transportability, etc. A trait that defines some aspect to produce quality is called quality trait. Each crop has a specific and often somewhat to completely different set of quality traits. Quality traits are classified as (1) morphological, (2) organoleptic, (3) nutritional, (4) biological, and (5) others.

Nutritional quality determines the value of the produce in human/animal nutrition. It includes carbohydrate content, protein content and quality, oil content and quality, vitamin content, mineral content, etc., and also the presence of antinutritional factors.

Like all living organisms, seeds are composed of many different types of chemicals, but seeds are unique in that they are a storehouse of chemicals that are used as food reserves for the next-generation plant. Seeds store three major classes of chemical compounds: carbohydrates (sugars), lipids (fats and oils), and proteins. These chemical foods also serve as a significant part of our food supply. The quantities of these quality compounds in seeds as well as growing parts vary with the genotype and environment of the crops.

Biofortification of various chemical compounds in grains is a new approach through conventional, molecular breeding or through agronomic manipulation. It holds considerable promise to increase nutritional status and health of poor population of developing world (Graham and Welch, 1966). Hence to breed or manipulate the existing genotypes for higher chemical compounds, phenotyping for biochemical traits is essential. In this context, protocols for quantifying different compounds are included in this section.

11.1 PROTEINS

Proteins are made of amino acids arranged in a linear chain joined together by peptide bonds. Many proteins are enzymes that catalyze the chemical reactions in metabolism. Other proteins have structural or mechanical functions, such as those that form the cytoskeleton, a system of scaffolding that maintains the cell shape. Some seeds do not have the optimum quantities of amino acids for human nutrition. For example, corn proteins are generally low in the amino acid lysine but relatively high

Phenotyping Crop Plants for Physiological and Biochemical Traits. http://dx.doi.org/10.1016/B978-0-12-804073-7.00011-9

in the amino acid methionine. In contrast, soybean proteins are relatively high in lysine but somewhat low in methionine. When corn and soybean seeds are used together, a nutritionally satisfactory balance can be obtained. Most seeds are poor in protein but are rich in carbohydrates or lipids. Soybean (high in protein and relatively high in lipids) is the exception rather than the rule.

11.1.1 PROTEIN ESTIMATION BY LOWRY METHOD

Protein can be estimated by different methods as described by Lowry and also by estimating the total nitrogen content. No method is 100% sensitive. Hydrolyzing the protein and estimating the amino acids alone will give the exact quantification. The method developed by Lowry et al. (1951) is sensitive enough to give a moderately constant value and hence is largely followed. Protein content of enzyme extracts is usually determined by this method.

Principle: Proteins react with Folin–Ciocalteau reagent to give a blue colored complex. The blue color developed is due to the reaction of the alkaline cupric tartarate with the protein and the reduction of the phosphomolybdic-phosphotungstic components in the Folin–Ciocalteau reagent by the amino acids tyrosine and tryptophan present in the protein. The blue color intensity is read at 660 nm using spectrophotometer.

Chemicals required

- Sodium carbonate (Na_2CO_3)
- Sodium hydroxide (NaOH)
- Copper sulfate ($CuSO4 \cdot 5H_2O$)
- Sodium potassium tartarate ($C_4H_4O_6KNa \cdot 4H_2O$)
- Folin–Ciocalteau reagent
- Trichloroacetic acid (CCl_3COOH)
- Bovine serum albumin

Preparation of reagents

- Reagent A: 2% sodium carbonate (Na_2CO_3 anhydrous): Dissolve 2 g of Na_2CO_3 in 0.1 N sodium hydroxide.
- Reagent B: 0.5% copper sulfate ($CuSO_4 \cdot 5H_2O$): 500 mg in 1.0% sodium potassium tartarate (prepare fresh).
- Reagent C: (alkaline copper solution): Mix 50 mL of solution A with 1.0 mL of solution B just before use.
- Folin–Ciocalteau reagent: Take 0.5 mL of Folin–Ciocalteau reagent and 0.5 mL D.D.H_2O for one sample.
- 2 N sodium hydroxide (NaOH): 8 g in 100 mL of D.D.H_2O
- 10% Tri chloro acetic acid (TCA): 10 g in 100 mL (50 mL TCA is cold and 50 mL TCA is normal).

Extraction:

1. Take 0.5 g sample and add 5 mL of 10% TCA (normal).
2. Grind the sample in a pestle and mortar.

3. The ground sample is taken in a centrifuge tube, add 5 mL of cold TCA, mix it, and keep it in a refrigerator for 15 min.
4. Centrifuge it at 3000 rpm for 15 min. Discard the supernatant.
5. Add 4 mL of 2 N NaOH to the pellet mix it and keep it for overnight.
6. Centrifuge it at 3000 rpm for 15 min.
7. Take 2 mL of supernatant and add 8 mL of distilled water.

Estimation:
1. Take 0.1 mL of sample solution and add distilled water to make up the volume to 1 mL and 1 mL of distilled water is added for blank.
2. Add 5 mL of reagent C to the sample solution, mix it, and keep for 10 min.
3. Add 0.5 mL of Folin–Ciocalteau reagent, mix well, and keep it for half an hour in dark.
4. Blue color will be developed.
5. Read the absorbance at 660 nm.
6. Take 50 mg of Bovine Serum Albumin (BSA) and dissolve in 50 mL of distilled water as stock.
7. Take 10 mL of stock and make up the volume to 50 mL with distilled water (1 mL contains 200 μg proteins).
8. Take 0.2, 0.4, 0.6, 0.8, and 1 mL solution from working standard of BSA, add distilled water, and make up to 1 mL.
9. Then proceed as that of the sample and read the color.

Calculation: Draw a standard curve using absorbance versus concentration. Find out the concentration of proteins in the sample using standard regression equation and express as mg per g fr.wt.

11.1.2 PROTEIN ESTIMATION BY BRADFORD METHOD (BRADFORD, 1976)

Principle: The method is based on the principle that the Coomassie Brilliant Blue (CBB) G-250 binds to proteins and gives a consist ant blue color. The protein dye complex has a higher extinction coefficient thus leading to a great sensitivity in measurement of protein. This binding of dye to protein is a very rapid process (approx. 2 min). The method is devoid of interferences by other soluble components.

Chemicals required:
- Monobasic sodium phosphate (NaH_2PO_4)
- Dibasic sodium phosphate (Na_2HPO_4)
- Bovine serum albumin
- 1 mM EDTA: 37.2 mg in 100 mL of D.D.H_2O
- 2% poly vinyl pyrrolidine
- Coomassie Brilliant Blue (CBB) G-250
- Ethanol
- Orthophosphoric acid

Preparation of reagents:
Phosphate buffer, 0.1M (pH: 7.8):

Solution A: 0.1 M Monobasic sodium phosphate (NaH_2PO_4)
Solution B: 0.1 M Dibasic sodium phosphate (Na_2HPO_4)
Add solution A of 8.5 mL to solution B of 91.5 mL and make up to a total volume of 200 mL with distilled water.

- 95% Ethanol: 95 mL of ethanol with 5 mL of D.D.H_2O
- 85% Orthophosphoric acid: 85 mL of OPA with 15 mL of D.D.H_2O

Bradford reagent:

CBB G-250 (3 mg) was dissolved in 1.5 mL of 95% Ethanol. To this solution 3 mL of 85% (W/V) Orthophosphoric acid was added. The resulting solution was diluted to a final volume of 30 mL with double distilled water, and freshly prepared reagent was used every time.

Extraction:

1. Take 1.0 g sample and grind to a thin paste and soluble proteins were extracted with 10 mL of phosphate buffer (pH: 7.8).
2. The extract was filtered through three layers of cheese cloth and centrifuged in cold (40°C) at 10,000 rpm for 10 min.

Estimation:

1. The supernatant of 100 μL was taken and to it 5 mL of Bradford reagent was added and mixed.
2. After 2 min and within 30 min.
3. Read the absorbance of solution at 595 nm.
4. Take 50 mg of BSA (Bovine Serum Albumin) and dissolved in 50 mL of distilled water as stock.
5. Take 10 mL of stock and make up the volume to 50 mL with distilled water (1 mL contains 200 μg proteins).
6. Take 0.2, 0.4, 0.6, 0.8, and 1 mL solution from working standard of BSA, add distilled water and make up to 1 mL.
7. Then proceed as that of the sample and read the color.

Calculation: Draw a standard curve using absorbance versus concentration. Find out the concentration of proteins in the sample using standard regression equation and express as mg per g fr.wt.

11.2 KERNEL OIL

11.2.1 OIL ESTIMATION BY SOXHLET APPARATUS (SOCS)

The fat or oil is extracted from the moisture free material with ether or petrol of boiling point below 60°C. Other substances waxes, resins, organic acids, coloring matter, etc., are also extracted with fat (Sadasivam and Manickam, 1992).

Chemicals required:

Petroleum ether
Soxhlet apparatus
Thimbles

Procedure:

1. Oven dry seed for 48 h if seed is fresh.
2. Otherwise dry for 30 min in oven at 100°C.
3. Grind 2 g of seed roughly.
4. Weigh 1 g of the ground material and transfer it to thimble.
5. Let the sample weight be W.
6. Mark the condensing flasks with the representative numbers and take its initial weight as W_1.
7. Then place the condensing flasks in the soxhlet apparatus.
8. Pour enough petroleum ether (80 mL) more than sufficient to run the cycle in the extractor.
9. Immerse the thimble into the extractor and place the extractor above the condensing flask.
10. Load all the condensing flasks along with extractors in the apparatus.
11. Connect the top of extractor to condenser.
12. Check the water level and connect instrument.
13. Open tap gently. Switch "ON" instrument.
14. Set the boiling point of solvent as boiling temperature.
15. The boiling temperature may be 10–20°C more than that of solvents boiling point. Example: boiling point of ether is 40–60°C. Boiling temperature can be 80°C.
16. Leave the process about 45–60 min.
17. After the process time, increase the temperature to recovery temperature (maximum boiling point × 2). Example: if the boiling point is 60°C, recovery temperature can be 120°C. Now do the rinsing about two times to collect the remaining oil that may be presented in the sample.
18. Now take out all the condensing flasks from the system and put them in a hot air oven at 80°C.
19. After 15–20 min, take out all the beakers and place them in a dessicator about 5 min.
20. Weigh the condensing flask as the final weight of beaker (W_2).
21. The amount of oil present in the sample can be calculated as

$$\%\text{oil} = \frac{W_2 - W_1}{W} \times 100$$

11.3 AFLATOXINS

11.3.1 QUANTIFICATION OF AFLATOXIN LEVELS IN KERNELS

Aflatoxins are highly carcinogenic, immunosuppressive agents, highly toxic and fatal to humans and particularly affecting liver and digestive tract. Katiyar et al. (2000) reported the risk of aflatoxins with hepatitis-B infection to human and

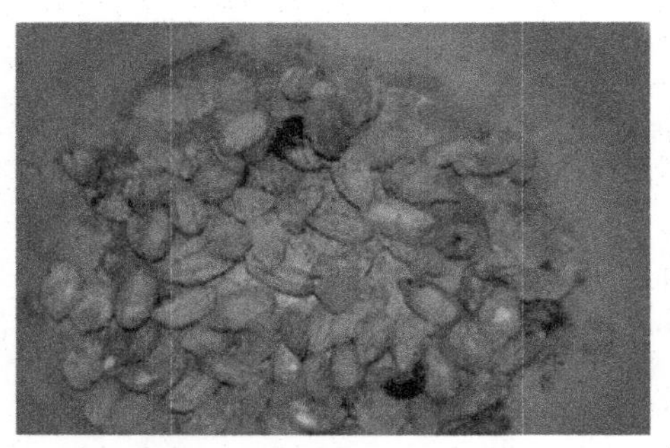

FIGURE 11.1 Aspergillus Flavus Infected Groundnut Kernels.

livestock population in India (*Aspergillus flavus* infected groundnut kernels shown in Fig. 11.1). It is currently known that there are synergistic effects between aflatoxin and Hepatitis-B infection causing liver cancer (Wogan, 1999). Among 18 types of aflatoxins reported, viz., B1, B2, G1, G2 (Fig. 11.2) are prominent and aflatoxin B1, (AfB1) is more carcinogenic and occur more commonly. M1 was reported in milk.

Aflatoxin is one of the major quality problems in grains of several crops, which hinders domestic consumption as well as the export potential, since the international regulation for minimum standards for aflatoxin contamination is becoming stringent. There were several reports of presence of higher levels of aflatoxin in the cultivated groundnut-based cattle and poultry feeds. In India aflatoxins are permissible up to 30 ppb food/grains and 120 ppb in poultry/animal feeds. In this laboratory quantification of aflatoxins by HPLC using Aflatest P columns and photochemical reactor for enhanced detection unit was standardized (Latha et al., 2011). The reliable and working protocol is given later:

Chemicals required:

HPLC methanol
HPLC water
Aflatest-p-column
Sodium chloride

Extraction of sample:

1. Weigh 20 g of sample into a blender jar.
2. Weigh 1 g of sodium chloride salt into blender jar.
3. Add 100 mL of 80% methanol.
4. Cap blender jar and seal with parafilm.
5. Blend at high speed for 1 min.
6. Filter the blender contents through a fluted filter paper into a 250 mL beaker.
7. Pipette 25 mL of filtrate into 50 mL graduated cylinder.
8. Add 25 mL of HPLC water to the cylinder.
9. Filter the content of the graduated cylinder through a glass fiber filter into a 250 mL beaker.
10. This filtrate will be used for passing through aflatest-p-column.

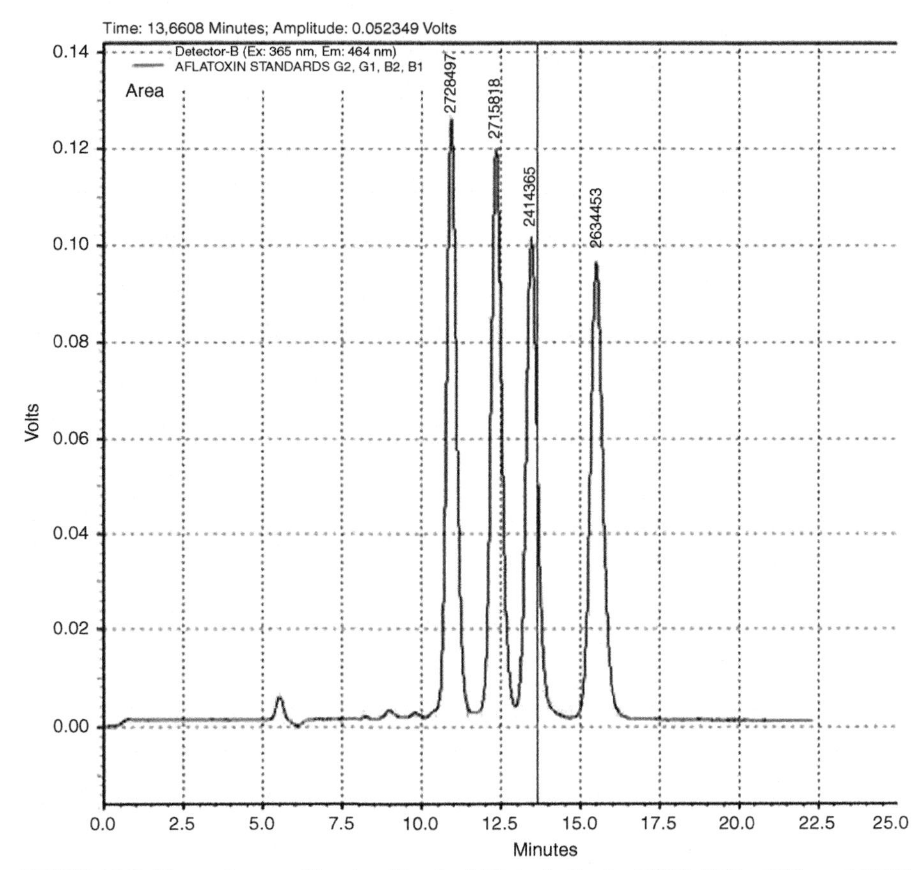

FIGURE 11.2 Chromatogram Showing Standard Aflatoxin Peaks (AfG2, AfG1, AfB2, and AfB1) Eluted at Particular Retention Times.

Immune-affinity column clean-up:

1. Pipette 10 mL of filtrate, pass through aflatest-p-column, and allow absorbing on column.
2. Once 10 mL of the sample has passed through the column, rinse the column with 10 mL of HPLC water.
3. Repeat HPLC water rinse for twice.
4. Add 1 mL of methanol to the column and collect all methanol eluent into a test tube.
5. The sample is now ready for injection into the HPLC.

HPLC instrument set-up protocol for estimation of aflatoxin:

1. Instrument set-up protocol for estimation of aflatoxins through HPLC in groundnut kernels was standardized.

2. Standardization was done by changing the mobile phase concentrations, flow rates, and injection volume of the standard to be injected.
3. Photochemical reactor for enhanced detection was used as postcolumn derivatization unit.
4. Finally, the chromatogram with four sharp aflatoxin peaks (AfB1, AfB2, AfG1, and AfG2) at particular retention time for standard is obtained.
5. The sample is now injected into the HPLC column for detection of aflatoxin in the sample. The retention time of the sample peaks should be similar to that of standard speaks.
6. The instrument set-up protocol for aflatoxin estimation is given later.
   a. Mobile phase: 63% HPLC water/0.02% nitric acid/22% HPLC methanol/15% HPLC acetonitrile
   b. Flow rate: 1.0 mL/min
   c. Column- symmetry: C-18
   d. Injection volume: 20 μL
   e. Detection: fluorescence detector having excitation at 365 nm and emission wavelength at 464 nm
   f. Gradient: isocratic
   g. Run time: 30 min

CHAPTER

Carbohydrates and related enzymes

12

Carbohydrates can be chemically divided into complex and simple. Simple carbohydrates consist of single or double sugar units (monosaccharides and disaccharides, respectively). Sucrose or table sugar (a disaccharide) is a common example of a simple carbohydrate. Complex carbohydrates contain three or more sugar units linked in a chain. They are digested by enzymes to release the simple sugars. Starch, for example, is a polymer of glucose units and is typically broken down to glucose. Simple and complex carbohydrates are digested at similar rates, so the distinction is not very useful for distinguishing nutritional quality. Carbohydrates are widely prevalent in the plants/grains of all crop plants, comprising the mono-, di-, oligo-, and polysaccharides. The common monosaccharides are glucose, fructose, galactose, lactose, and maltose. Protocols for sugars and related enzymes are given in this chapter which can be used for estimation in plant parts or kernels.

12.1 REDUCING SUGARS

Sugars with reducing property (arising out of the presence of a potential aldehyde or keto group) are called reducing sugars. Some of the reducing sugars are glucose, galactose, lactose, and maltose. Estimation of reducing sugars using dinitrosalicylic acid method is simple, sensitive, and adoptable during handling of a large number of samples at a time (Krishnaveni et al., 1984; Miller, 1972).

Chemicals required:

- Dinitrosalicylic acid
- Ethanol
- Sodium hydroxide
- Sodium sulfite
- Crystalline phenol
- Potassium sodium tartrate (Rochelle salt)
- Glucose

Preparation of reagent:

- *Di nitrosalicylic acid reagent (DNS):* dissolve by stirring 1 g dinitrosalicylic acid, 200 mg crystalline phenol, and 50 mg sodium sulfite in 100 mL of 1% sodium hydroxide. Store at 4°C. Prepare fresh before use.

Phenotyping Crop Plants for Physiological and Biochemical Traits. http://dx.doi.org/10.1016/B978-0-12-804073-7.00012-0

Extraction:

1. Weigh 100 mg of the sample and extract the sugars with hot 80% ethanol twice (5 mL each time).
2. Collect the supernatant and evaporate it by keeping it on a water bath at 80°C.
3. Add 10 mL water and dissolve the sugars.

Estimation:

1. Take 0.5–3.0 mL of extract for analysis in test tube and equalize the volume to 3 mL with water in all the test tubes.
2. Add 3 mL of DNS reagent. Heat the contents in a boiling water bath for 5 min.
3. When the contents of the tubes are still warm, add 1 mL of 40% Rochelle salt solution.
4. Cool and read the dark red color intensity in a spectrophotometer using photometric method at 510 nm.
5. Prepare the reagent blank as above by taking 1 mL of distilled water instead of the extract.
6. Dissolve 100 mg glucose in 100 mL of distilled water in a volumetric flask and taken as stock.
7. Take 10 mL of this stock standard and dilute to 100 mL in another volumetric flask for working standard solution.
8. A series of volume from 0.0 to 1.0 mL of this standard solution gives a concentration range 10–100 μg.
9. Then proceed as that of the sample and read the color.

Calculation: Draw a standard curve using absorbance versus concentration. Find out the concentration of the reducing sugars in the sample using standard regression equation and express as "mg per g fresh wt."

12.2 NONREDUCING SUGARS

The content of nonreducing sugars can also be calculated by subtracting the reducing sugars from total carbohydrate content.

12.3 TOTAL CARBOHYDRATES

Carbohydrates are the important components of storage and structural materials in plants. They exist as free sugars and polysaccharides. The basic units of carbohydrates are monosaccharides which cannot be split by hydrolysis into more simple sugars. The carbohydrate content can be measured by hydrolyzing the polysaccharides into simple sugars by acid hydrolysis and estimating the resultant monosaccharides.

Principle: Carbohydrates are first hydrolysed into simple sugars using dilute hydrochloric acid. In hot acidic medium glucose is dehydrated to hydroxymethyl furfural. This compound condenses with anthrone to form blue-green colored product that is measured at 630 nm (Hedge and Hofreiter, 1962).

Chemicals required:

- Hydrochloric acid (HCl)
- Sulfuric acid (H_2SO_4)
- Anthrone
- Glucose

Preparation of reagents:

- 2.5 N hydrochloric acid (HCl): 20.83 mL of 35% HCl make up to 100 mL of D.D.H_2O.
- Anthrone reagent: dissolve 200 mg anthrone in 100 mL of ice cold 95% H_2SO_4. Prepare fresh before use.

Extraction:

1. Weigh 100 mg of sample into a test tube.
2. Add 5 mL of 2.5 N HCl. The test tube is then kept in hot water bath for 3 h and cool to room temperature.
3. Add solid sodium carbonate until the effervescence ceases to neutralize it.
4. Make up the volume to 50 mL with distilled water.
5. Centrifuge for 20 min at 2000 rpm.
6. Collect the supernatant.

Estimation:

1. Take 1 mL of aliquot for analysis.
2. Cool the aliquot on ice and add 4 mL of ice cold anthrone reagent.
3. Keep in hot water bath for 8 min.
4. Cool rapidly and read the green color intensity in a spectrophotometer using photometric method at 630 nm.
5. Prepare the reagent blank as above by taking 1 mL of distilled water instead of the extract.
6. Dissolve 100 mg glucose in 100 mL of distilled water in a volumetric flask and taken as stock.
7. Take 10 mL of this stock standard and dilute to 100 mL in another volumetric flask for working standard solution.
8. A series of volumes from 0.0 to 1.0 mL of this standard solution gives a concentration range 10–100 μg.
9. Then proceed as that of the sample and read the color.

Calculation: Draw a standard curve using absorbance versus concentration. Find out the concentration of the total carbohydrates in the sample using standard regression equation and express as milligram per gram fresh wt.

12.4 ESTIMATION OF SUCROSE PHOSPHATE SYNTHASE

Sucrose phosphate synthase is estimated following Huber and Bickoff's (1984) method as given later:

Chemicals required:

- Magnesium chloride ($MgCl_2$)
- Ethylene di amine tetra acetic acid (EDTA)
- Dithiothreitol (DTT)
- Bovine serum albumin (BSA)
- UDP-glucose
- Fructose-6-phosphate
- Hepes-NaOH buffer
- Potassium hydroxide (KOH)

Preparation of reagents:

- 5 mM magnesium chloride ($MgCl_2$)
- EDTA (1 mM): 37.2 mg in 100 mL of D.D.H_2O
- 2.52 mM dithiothreitol (DTT)
- 0.5% BSA
- 7.5 mM UDP-glucose
- 7.5 mM fructose-6-phosphate
- 50 mM, pH 7.5 Hepes-NaOH buffer
- KOH (30% solution): 30 g in 100 mL of D.D.H_2O

Extraction:

1. Take 1 g of fresh plant material, homogenize with 10 mL of grinding medium [consisting of 50 mM Hepes-NaOH buffer (pH 7.5), 5 mM $MgCl_2$, 1 mM Na-EDTA, 2.5 mM DTT and 0.5% BSA].
2. Filter the slurry through a 4-layered cheese cloth.
3. Centrifuge at 38,000× g for 10 min at 0–4°C.
4. Collect the supernatant and preserve it for enzyme assay.

Estimation:

1. Sucrose phosphate synthase is assayed by the measurement of fructose-6-phosphate-dependent formation of sucrose from UDP-glucose (Huber and Bickoff, 1984).
2. The assay can be initiated by addition of 20 μL of enzyme extract to a reaction mixture containing 50 mM Hepes-NaOH buffer (pH 7.5), 5 mM $MgCl_2$, 7.5 mM UDP-glucose; and 7.5 mM fructose-6-phosphate in a total volume of 0.1 mL.
3. The reaction is performed in test tube at 30°C using a shaking water bath and terminated after 10 min by adding 0.1 mL of 30% KOH solution. (Background is determined by including zero reaction time samples).
4. The tubes are subsequently kept in a boiling water bath for 10 min to destroy the unreacted fructose-6-phosphate.
5. After cooling, OD is recorded at 620 nm and sucrose content is determined following anthrone method.
6. Sucrose phosphate synthase activity is expressed as "μmol sucrose g^{-1} fr.wt.h^{-1}."

12.5 ESTIMATION OF STARCH SYNTHASE

Leloir and Goldenberg's (1960) method can be conveniently used for estimation of the enzyme "starch synthase." The enzyme occurs in soluble form [soluble starch synthase (SSS)] or granule-bound form (granule bound starch synthase).

Principle: The reversible reaction catalyzed by sucrose synthase as given here:

$$\text{Sucrose} + \text{UPP} \rightleftharpoons \text{UDP} - \text{glucose} + \text{fructose}$$

Chemicals required:

- Tris-hydroxymethyl aminomethane
- Hydrochloric acid (HCl)
- Magnesium chloride ($MgCl_{2)}$
- Dithiothreitol (DTT)
- Bovine serum albumin (BSA)
- Ethylene diamine tetra acetic acid (EDTA)
- Glutathione
- Potassium chloride (KCl)
- Glycogen
- ADPG

Preparation of reagents:
Grinding medium:

- Tris-HCl Buffer: 50 mM; pH 7.5
- $MgCl_2$: 5 mM
- Dithiothreitol (DTT): 1 mM
- Bovine serum albumin (BSA): 1 mg/mL

Assay medium:

- Tris-HCl Buffer: 25 μmol; pH 8.5
- EDTA: 0.2 μmol
- Glutathione: 2.5 μmol
- KCl: 5 nmol
- Glycogen: 0.5 mg
- ADPG: 0.25 μmol

Extraction:

1. Plant sample of 1 g is homogenized in a prechilled pestle-mortar with 10 mL of grinding medium.
2. The homogenate is centrifuged at 30,000× g for 20 min at 0–4°C.
3. The supernatant is stored for assay of soluble starch synthase.
4. Furthermore, the pellet is suspended in 10 mL of grinding medium and centrifuged at 30,000× g for 20 min at 0–4°C.
5. The pellet so obtained is again washed with 5 mL of the grinding medium and centrifuged again.

6. The pellet finally obtained is resuspended in 1 mL of the same buffer to obtain globule-bound starch synthase activity (George et al., 1994).

Estimation:

1. In a test tube, take 0.1 mL of the enzyme extract (supernatant) and add 0.25 mL of reaction mixture containing (25 mM Tris-HCl buffer, pH 8.5, 0.2 μM EDTA, 2.5 μM glutathione, 5 μM KC1, 0.5 mg glycogen, and 0.25 μM ADPG).
2. The reaction is started by adding the enzyme extract and stopped later by immersing the tube in a boiling water bath for 1 min.
3. The tubes are cooled immediately in a water bath at room temperature.
4. A blank is run without the substrate ADPG and amount of ADP liberated is then determined by pyruvate kinase method (Leloir and Goldenburg, 1960).
5. To estimate the ADP formed in the above reaction, 0.025 mL of phosphoenol pyruvate (0.01 M solutions in 0.4 M KC1) and 0.025 mL of pyruvate kinase (8.4 U freshly diluted in 0.1 M $MgSO_4$) are added and incubated for 15 min at 37°C.
6. At the end of the incubation period, 0.15 mL of dinitrophenyl hydrazine (0.1% in 2 N-HC1) is added.
7. After 5 min, 0.2 mL of 10 N NaOH and 1.1 mL of 95% ethanol are added.
8. The samples are mixed and centrifuged and the OD of the supernatant is measured at 520 nm.
9. Results are calculated from a standard curve drawn by using different concentrations (10–100 nmol) of pyruvate.
10. The enzyme activity is expressed on fr.wt. or dry weight basis.

12.6 ESTIMATION OF INVERTASES

1. Invertases are prevalent in storage tissues and can catalyze the hydrolysis of sucrose to glucose and fructose.
2. This reaction is termed "inversion reaction" because during the reaction the optical rotation shifts from +66.5°C to –28°C.
3. Invertase is also called "sucrose."
4. Since the hydrolysis is not reversible, sucrase cannot catalyze the synthesis of sucrose.

Principle: The enzyme sucrose phosphate synthase catalyzes the synthesis of sucrose and the enzymes invertase and sucrose synthetase both catalyze the hydrolysis of sucrose, the later producing the nucleotide diphosphate derivatives of glucose for subsequent biosynthetic reactions.

The enzyme invertase can be quantitatively assayed following a method as described by Tang et al. (1999).

Chemicals required:

- Sodium acetate ($C_2H_3O_2Na.3H_2O$)
- Acetic acid
- β-Mercaptoethanol
- Lysine
- Ethylene diamine tetra acetic acid (EDTA)
- Phenylmethane sulfonyl fluoride.
- Sodium chloride (NaCl)
- Sucrose
- Citric acid
- Sodium carbonate (Na_2CO_3)
- Rochelle salt–sodium potassium tartarate
- Copper sulfate
- Sodium bicarbonate ($NaHC^{O}{}_3$)
- Sodium sulfate (Na_2SO_4)

Preparation of reagents:
Sodium acetate buffer (25 mM; pH 5.0)

A: 0.5% β-mercaptoethanol (500 μL in 100 mL) + 25 mM solution of acetic acid (1.444 mL in 1000 mL D.D H_2O water).

B: 25 mM solution of sodium acetate ($C_2H_3O_2Na{\cdot}3H_2O$): 0.3402 g of sodium acetate in 100 mL of D.D H_2O water).

14.8 mL of A + 35.2 mL of B, diluted to a total of 100 mL. Adjust the pH (5.0)

- Lysine (10 mM)
- EDTA (1 mM) : 37.2 mg in 100 mL of D.D.H_2O
- 0.1 mM phenyl methane sulfonyl fluoride.
- NaCl (1 M): 5.844 g in 100 mL of D.D.H_2O water.

Cell wall invertase:

- Sucrose (50 mM)
- Citric acid (13.5 mM): 259.36 mg in 100 mL of D.D.H_2O.
- Disodium phosphate, 26.5 mM (pH 4.6): 376.19 mg in 100 mL D.D.H_2O.

Soluble invertase:

- Citric acid (10.5 mM): 201.72 mg in 100 mL of D.D.H_2O
- Disodium phosphate (29.0 mM) (pH 5.4): 411.68 mg in 100 mL of D.D.H_2O
- Alkaline copper carbonate tartrate reagent according to Somogyi (1952).

Extraction:

1. Leaves (2 g) of mature plants are homogenized in 10 mL of ice-cold sodium acetate buffer (25 mM; pH 5.0) containing 0.5% β-mercaptoethanol, 10 mM lysine, and 1 mM EDTA and 0.1 mM phenylmethane sulfonyl fluoride.
2. The homogenate is filtered through a 4-layered muslin cloth.
3. It is centrifuged at 10,000× g for 30 min at 0.4°C.

4. The supernatant is used for the determination of acid soluble invertase activity.
5. The pellets are washed extensively with ice-cold water and cell wall proteins are extracted with 5 mL of 1 M NaCl overnight at 4°C.

Estimation:

Cell wall invertase: The assay mixture consists of 50 mM sucrose, 13.5 mM citric acid, 26.5 mM disodium phosphate (pH 4.6), and pellet extract all in a total volume of 3.0 mL.

Soluble invertase: Constituents of assay mixture are 50 mM sucrose, 10.5 mM citric acid, 29.0 mM disodium phosphate (pH 5.4), and the supernatant all in a total volume of 3.0 mL.

The reaction mixture is incubated at 37°C for 15–20 min.

The reaction is stopped with alkaline copper reagent and the liberated reducing sugars are measured following the method of Somogyi (1952).

CHAPTER

Nitrogen compounds and related enzymes

13

Plants take up nitrogen from soil either in the form of nitrate or ammonia. Nitrate taken by plants is reduced by nitrate reductase and nitrite reductase enzymes. They catalyze step-wise reduction of nitrate to nitrite and nitrite to ammonia. Ammonia occupies the key position in nitrogen metabolism for the synthesis of organic nitrogen. Glutamic acid dehyrogenase (GDH), glutamate synthase (GOGAT), and glutamine synthatase (GS) are important enzymes that are involved in the conversion of ammonia into glutamic acid, a primary amino acid. The α-amino group of several amino acids is formed as a result of amino transferase enzymes. This process is of great significance in amino acid metabolism. Protocols for estimation of all the important enzymes associated with nitrogen metabolism as well as total nitrogen and free amino acids are given in this chapter.

13.1 TOTAL NITROGEN

13.1.1 KJELDHAL METHOD FOR QUANTIFYING LEAF NITROGEN CONTENT

Measurement of leaf nitrogen can be made using the Kjeldhal method. The Kjeldhal method for quantifying leaf nitrogen status involves digesting a dried and finely powered sample in a medium containing a strong acid such as sulfuric acid to produce ammonium sulfate, followed by liberation of ammonia by adding a strong alkali–sodium hydroxide. The ammonia is then captured using boric acid and the exact amount of nitrogen can be determined by titrating the excess acid with sodium carbonate. This method is invariably followed in this laboratory for computing specific leaf nitrogen in crop genotypes. Although the Kjeldhal method is fairly accurate, it is quite cumbersome and time-consuming.

Principle:

A known weight of the powdered sample is treated with sulfuric acid so as to oxidize the organic matter and bring the mineral elements into solution.

Chemicals required:

Sulfuric acid
Potassium sulfate
Copper sulfate ($CuSO_4 \cdot 5H_2O$)
Boric acid

Phenotyping Crop Plants for Physiological and Biochemical Traits. http://dx.doi.org/10.1016/B978-0-12-804073-7.00013-2

Bromocresol green
Methyl red double indicator
Sodium hydroxide

13.1.2 PREPARATION OF REAGENTS

Receiver solution: Receiver solution of 1 L was prepared by adding 40 g of boric acid, 10 mL of bromocresol green, 7 mL of methyl red solution into a conical flask and making up to 1 L with distilled water.

40% sodium hydroxide: 40 g of sodium hydroxide in 100 mL of distilled water
N sulfuric acid: 2.7 mL of sulfuric acid in 1 L of distilled water

Digestion: Transfer 1 g of powdered sample into digestion tube. Add 0.8 g of copper sulfate and 7 g of potassium sulfate to the digestion tube. Concentrated sulfuric acid (12 mL) was added slowly to the digestion tube. The digestion tube was kept on a digestion plate and heat at 420°C for about 30–45 min. Continue to digest until the sample turns to blue/green color.

Distillation: Remove the digestion tubes and cool the tubes for 10–20 min. Add 75 mL of deionised water to digestion tube and keep it in distillation unit. Add 25 mL of receiver solution into 250-mL conical flask and place it into distillation unit. Raise the platform so that the distillate outlet is submerged in the receiver solution. Dispense 50 mL of 40% sodium hydroxide into the digestion tube. Open the steam valve and distill approximately for 4 min. The receiver solution in the distillation unit will turn to green color indicating the presence of alkali ammonia. Take out the flask having receiver solution and titrate against 0.1 N sulfuric acid.

Calculation:

$$\text{Percent nitrogen (\% N)} = \frac{(T - B) \times N \times 14.007 \times 100}{\text{Weigth of sample in mg}}$$

where

T = titre value
B = blank value
N = normality of sulfuric acid (0.1 N)

13.1.3 PROTEIN PERCENT CAN BE DETERMINED INDIRECTLY USING THE FOLLOWING FORMULA

$$\text{Percent protein} = 6.25 \times \%\ \text{N (A.O.A.C, 1960)}$$

Other methods: More sophisticated instruments such as elemental analyzers provide a high-throughput measurement option for accurate determination of leaf nitrogen. In this technique, a small quantity of the homogenized sample is completely combusted in a temperature-controlled reactor filled with appropriate catalysts. In the presence of excess oxygen for combustion, the nitrogen in the sample is oxidized to produce nitrogen oxides. These nitrogen oxides are subsequently reduced to form nitrogen gas. The quantity of nitrogen gas is determined by a temperature conductivity detector.

13.2 TOTAL FREE AMINO ACIDS

The amino acids are the organic compounds that form the basic building blocks of proteins. The common feature of the structure of amino acids is having a minimum of two ionizable groups: the acidic carboxyl (-COOH) and the basic amino (-NH2) groups on the same carbon atom called α-carbon atom. The amino acids that also exist in the free form and are not bound to proteins are known as free amino acids. They are mostly water soluble in nature. Very often in plant during disease conditions, the free amino acids composition exhibits a change and hence, the measurement of the total free amino acids gives the physiological and health status of the plants.

Principle: Ninhydrin, a powerful oxidizing agent, decarboxylates the alpha-amino acids and carboxylates to give an intensely colored bluish purple product that is colorimetrically measured at 570 nm (Moore and Stein, 1948; Misra et al., 1975; Theymoli Balasubramanian and Sadasivam, 1987).

$$\text{Ninhydrin} + \alpha\text{-amino acid} \rightarrow \text{Hydrindantin} + \text{decarboxylated Amino acid} + \text{carbon dioxide} + \text{Ammonia.}$$

$$\text{Hydrindantin} + \text{Ninhydrin} + \text{Ammonia} \rightarrow \text{Purple colored product} + \text{Water}$$

Chemicals required:

- Ninhydrin
- Stannous chloride
- Citric acid
- Sodium citrate
- Methyl cellosolve
- N-propanol
- Ethanol
- Leucine

Preparation of reagents:

- Ninhydrin: dissolve 0.4 g stannous chloride ($SnCl_2 \cdot 2H_2O$) in 250 mL of 0.2 M citrate buffer (pH 5.0). Add this solution to 10 g of ninhydrin in 250 mL of methyl cellosolve (2-methoxyethanol). Prepare freshly and store in brown bottle (carcinogenic).
- 0.2 M citrate buffer (pH 5.0)
- Solution A: 0.2 M citric acid solution
- Solution B: 0.2 M sodium citrate ($C_6H_5O_7\ Na_3 \cdot 2H_2O$) solution

Add 20.5 mL of solution A and 29.5 mL of solution B diluted to a total of 100 mL with distilled water (pH 5.0).

Diluent solvent: mix equal volumes of water and n-propanol, and use.

Extraction:

1. Weigh 500 mg of the plant sample and grind it in a pestle and mortar with 5–10 mL of 80% ethanol.
2. Filter or centrifuge. Save the filtrate or the supernatant.
3. Repeat the extraction twice with residue and pool all the supernatants.
4. Reduce the volume if needed by evaporation and use the extract for the quantitative estimation of total free amino acids.
5. If the tissue is tough, use boiling 80% ethanol for extraction.

Estimation:

1. To 0.1 mL of extract add 1 mL of ninhydrin solution and mix.
2. Make up the volume to 2 mL with distilled water.
3. Heat the tube in a boiling water bath for 20 min.
4. Add 5 mL of the diluents and mix the contents.
5. After 15 min, read intensity of purple color against a reagent blank in a spectrophometer using photometric method at 570 nm.
6. The color is stable for 1 h. Prepare the reagent blank as above by taking 0.1 mL of 80% ethanol instead of the extract.
7. Dissolve 50 mg leucine in 50 mL of distilled water in a volumetric flask.
8. Take 10 mL of this stock standard and dilute to 100 mL with distilled water in another volumetric flask for working standard solution.
9. A series of volumes from 0.1 to 1.0 mL of this standard solution gives a concentration range 10–100 μg.
10. Then proceed as that of the sample and read the color.

Calculation: Draw a standard curve using absorbance versus concentration. Find out the concentration of the free amino acids in the sample using standard regression equation and express as μg per g fr.wt.

13.3 NITRATE REDUCTASE

The assimilatory reduction of nitrate by plant is a fundamental biological process in which a highly oxidized form of inorganic nitrogen is reduced to nitrite and then to ammonia.

$$NO_3^- + AH_2 \rightarrow NO_2^- + A + H_2O$$

The nitrate reducing system consists of nitrate reductase and nitrite reductase which catalyze stepwise reduction of nitrate to nitrite and then to ammonia. The NADH-dependent NR is most prevalent in plants.

Principle: Nitrate reductase (NR) is capable of utilizing the reduced form of pyridine nucleotides, flavins, or benzyl viologen as electron donors for reduction of nitrate to nitrite. Here, NR activity in plants can be measured by following the oxidation of NAD (P) H at 340 nm. However, NR activity is commonly measured by spectrophometric determination of nitrite produced (Klepper et al., 1971).

Chemicals required:
- Dipotassium hydrogen ortho phosphate (K_2HPO_4)
- Potassium di hydrogen ortho phosphate (KH_2PO_4)
- Potassium nitrate (KNO_3)
- Sulphanilamide
- 1-Naphthyl ethylene diamine dihydrochloride (NEDD)
- N-propanol
- Sodium nitrite ($NaNO_2$)

Preparation of reagents:

Potassium phosphate buffer (0.05M), pH 7.5

Solution A: Dipotassium hydrogen ortho phosphate (K_2HPO_4) 0.8709 g in 100 mL of D.D.H_2O

Solution B: Potassium di hydrogen ortho phosphate (KH_2PO_4) 0.68045 g in 100 mL of D.D.H_2O

Add 70 mL of solution A to 30 mL of solution B and adjust the pH to 7.5 with NaOH.

- Substrate solution (0.4 M): 4.011 g of potassium nitrate (KNO_3) in 100 mL of distilled water
- Sulphanilamide (1.0%): 1.0 g of sulphanilamide in 100 mL of 2.4 N HCl
- NEDD (0.02%): 0.02 g in 100 mL of distilled water

Extraction:
1. Weigh 200 mg of leaf sample, cut into small pieces, and take into 15-mL culture tubes.
2. To this add 2 mL of 0.1M chilled phosphate buffer, 1.0 mL of 0.1M chilled and freshly prepared potassium nitrate solution, and 0.1 mL of N-propanol.
3. These tubes were kept for 1 h at room temperature and incubated at 30°C for 30 min. Then these tubes were placed at 100°C for 2 min.
4. Cool the culture tubes and decant the solution (filtrate).

Estimation:
1. Take fresh test tubes for estimation and filled with 1.0 mL of 1.0% sulphanilamide followed by 1.0 mL of 0.02% of NEDD solution.
2. Add 0.2 mL of filtrate and finally make up the volume with 8 mL of distilled water.
3. Wait for 15–20 min for pink color formation.
4. Read the absorbance of color intensity at 540 nm in UV–VIS spectrophotometer using photometric method mode.
5. Stock solution is prepared by weighing 69 g of sodium nitrite ($NaNO_2$) in 1000 mL of distilled water (1M) or 69 mg of sodium nitrite in 1000 mL of sodium nitrite (1 mM).
6. Take 10 mL of stock solution and make up the volume to 100 mL for working standard.
7. Take ten different concentrations (0.1, 0.2, 0.3, 0.4, 0.5, 0.6, 0.7, 0.8, 0.9, 1.0 mL) of $NaNO_2$ and then proceed as that of the sample and read the color at 540 nm.

Calculation: Express the NR activity of nitrate reductase as μmoles $NaNO_2$ formed h^{-1} g^{-1} fresh wt. or per minute per milligram protein.

13.4 NITRITE REDUCTASE

Nitrite is directly reduced to ammonia by nitrite reductase without the liberation of free intermediates.

$$NO_2^- \xrightarrow[+6e]{\text{Nitrite reductase}} NH_4^+$$

The enzyme accepts electrons from photosynthetically reduced ferredoxin but not from reduced pyridine nucleotides (Vega et al., 1980).

Chemicals required:

- Tris-hydroxymethyl amino methane
- Hydrochloric acid (HCl)
- Sodium nitrite solution
- Methyl viologen solution
- Sodium dithionite ($Na_2S_2O_4$)
- Sodium bicarbonate solution ($NaHCO_3$) (2.5%)

Preparation of reagents:

- Tris-HCl buffer 0.5 M (pH 7.5)
- Sodium nitrite solution ($NaNO_2$): dissolve 43.2 mg $NaNO_2$ in 20-mL distilled water.
- Methyl viologen solution: dissolve 60.1 mg methyl viologen in 20-mL water.
- Sodium dithionite-bicarbonate solution: dissolve 250 mg each of $Na_2S_2O_4$ (2.5%) and $NaHCO_3$ (2.5%) in 10-mL water.

Extraction:

1. Homogenize the leaf tissue (10 g/100 mL) with Tris-HCl buffer (pH 7.5) in a warring blender at high speed for 3 min and filter through eight layers of cheese cloth at 4°C. Use the filtrate as an enzyme source.

Estimation:

1. Prepare a reaction mixture by mixing 6.25 mL of Tris-HCl buffer, 2 mL of sodium nitrite solution, 2 mL methyl viologen solution, and 14.75 mL water. Pipette out 1.5 mL reaction mixture and 0.3 mL of enzyme preparation into a test tube. Run a blank without enzyme.
2. Start the reaction by adding 0.2 mL of freshly prepared dithionite-sodium bicarbonate solution.
3. Incubate for 15 min at 30°C.
4. Shake the tube vigorously (use vortex mixer) until blue color disappears.

5. Take 20 μL of the aliquot in a test tube; add 1 mL each of sulphanilamide and NEDD into it. Wait for 30 min.
6. Measure the absorbance at 540 nm in a colorimeter.

Calculation: The enzyme activity is expressed as the amount of nitrite (μM) reduced per min per mg protein.

13.5 LEGHEMOGLOBIN (Lb)

1. Leghaemoglobin is found in the nodules of leguminous plants.
2. The main functions of leghemoglobin are (1) to facilitate oxygen supply to the nitrogen fixing bacteria and (2) to protect the enzyme, nitrogenase from being inactivated by oxygen.
3. Hence, the presence of leghaemoglobin exhibits a good coordination between host plant and the bacteria.

Principle: Leghaemoglobin content may be assayed following the method of Hartree (1955). Leghaemoglobin estimation is based on the fact that it forms greenish-yellow-colored hemochrome, when it reacts with pyridine in an alkaline medium. The color concentration measured at 556 nm gives an estimate of leghaemoglobin.

Chemicals required:

- Mono basic sodium phosphate (NaH_2PO_4)
- Dibasic sodium phosphate (Na_2HPO_4)
- Sodium hydroxide (NaOH)
- Pyridine
- Sodium dithionite ($Na_2S_2O_4$)

Preparation of reagents

Phosphate buffer (0.1 M; pH 7.0)

A. 1.1998 g of NaH_2PO_4 in100 mL of D.D.H_2O.

B. 1.4196 g of Na_2HPO_4 in 100 mL of D.D.H_2O.

39 mL of A + 61 mL of B diluted to a total of 200 mL. Adjust pH 7.0 before dilution.

Sodium hydroxide (NaOH) solution (0.2 M)): 0.4 g in 100 mL of D.D.H_2O.

Extraction:

1. Take 1–2 g fresh or thawed nodules.
2. Grind in 10–20 mL of chilled phosphate buffer with pestle.
3. And mortar kept in an ice-bath or tray.
4. Filter the macerate through 2–4 layered muslin cloth.
5. Centrifuge the filtrate at 10,000 × g for 20 min at 4°C.
6. Discard the pellets.
7. Take the supernatant in a test tube and make the volume up to 5 mL with phosphate buffer.

Estimation:

1. Take the supernatant in a test tube.
2. Add equal volume of 0.2 M NaOH and leave it for 30 min.
3. Add a pinch of $Na_2S_2O_4$ (about 100 mg) and keep the tube at room temperature for an hour.
4. Centrifuge the contents of the tube at 5000 rpm for 20 min at 4°C.
5. Take the aliquot and measure its absorbance at 556 nm against a reagent blank, using a colorimeter or spectrophotometer.
6. Calculate the leghaemoglobin content using the standard curve of graded concentrations of haemoglobin (M/s Sigma) and express the values as milligram per gram fresh weight of nodules.
7. For best results 30–40-day old nodules should be taken for assay of leghaemoglobin.
8. Nodules can be picked into liquid nitrogen or chilled phosphate buffer.
9. Nodules can be stored frozen, but it is recommended not to store for a long period.

13.6 GLUTAMIC ACID DEHYDROGENASE (GDH)

Chemicals required:

- Tris-hydroxymethyl aminomethane
- Hydrochloric acid (HCl)
- α-Ketoglutaric acid
- Ammonium sulfate
- Nicotinamide adenine dinucleotide reduced (NADH)

Preparation of reagents:

- Tris-HCl buffer; 0.2 M (pH 8.2)
- α-ketoglutaric acid (20 μM)
- Ammonium sulfate (150 μM) in distilled water
- Nicotinamide adenine dinucleotide reduced (NADH): 0.2 μM in Tris-HCl buffer.

Extraction:

1. Sample (20 g) is placed in a precooled mortar.
2. It is ground thoroughly to a paste with equal volume of Tris-HCl buffer (pH 8.2).
3. The crude homogenate is passed through four layers of muslin cloth and the extract is centrifuged at 16,000 rpm (22,000 × g) for 30 min at –4°C in a refrigerated centrifuge.
4. The supernatant is saturated to 40% with ammonium sulfate and centrifuged at 10,000 rpm (16,000 × g) for 15 min.
5. Supernatant is taken and precipitate discarded.
6. The supernatant is saturated to 60% with ammonium sulfate and centrifuged after a period of 20 min, at 10,000 rpm (16,000 × g) for 15 min.

7. The resultant precipitate (between 40% and 60%) is dissolved in 5 mL of Tris-HCl buffer (0.2 M; pH 8.2) at 5°C for 18 h with a continuous stirring.
8. Cold buffer should be replaced three to four times during dialysis.

Estimation:

1. Glutamic acid dehydrogenase activity is assayed following the oxidation of NADH and measured spectrophotometrically at 340-nm wavelength (Bullen, 1956).
2. The reaction mixture comprises the following:
   a. 0.1 mL enzyme extract
   b. 0.1 mL α-ketoglutaric acid (20 μM)
   c. 0.1 mL ammonium sulfate (150 μM)
   d. 0.2 mL NADH (0.2 μM)
   e. 2.5 mL Tris-HCl buffer (0.2 M, pH 8.2)
3. Final volume of the reaction mixture is made to 3 mL in a cuvette by adding buffer.
4. A blank with all the substrates except NADH is used as control.
5. The optical density is adjusted to a point and the decrease in absorbency per minute is recorded continuously for 10 min.

Calculation: Specific activity of the enzyme is expressed as micromoles of NADH oxidized per minute per milligram of soluble enzyme protein.

13.7 GLUTAMATE SYNTHASE (GOGAT)

The original name given to this enzyme was glutamine (amide)-2-oxoglutarate amino transferase (oxidoreductase $NADP^+$) from which the acronym GOGAT is derived. The trivial name glutamate synthase is also very much in use. Glutamate synthase is assayed spectrophotometrically by recording the rate of oxidation of NADPH or NADH, as indicated by a change in absorbance at 340 nm following the addition of enzyme extract (Tempest et al., 1970).

Chemicals required:

- Tris-hydroxymethyl aminomethane
- Hydrochloric acid (HCl)
- Glutamine
- 2-Oxoglutarate
- NADPH
- Disodium EDTA
- Dithiothreitol (DTT)
- Poly vinyl pyrrolidine (PVP)

Preparation of reagents:

- Tris-HCl buffer; 50 mM (pH 7.6)
- Preparation of the following reagents in Tris HCl buffer, 50 mM (pH 7.6)

- Glutamine: 5 mM (36.5 mg/10 mL)
- 2-Oxoglutarate: 5 mM (36.5 mg/10 mL)
- NADPH: 0.25 mM (10 mg/10 mL)
- Disodium EDTA (1 mM): (3.7224 mg in 10 mL)
- Dithiothreitol, 1 mM (DTT) (15.43 mg per 100 mL)
- 1% poly vinyl pyrrolidine (PVP): (100 mg in 10 mL)

Extraction:

1. Grind 1 g of the plant material with 5 mL of 100 mM phosphate buffer (pH 7.5) containing 1 mM disodium EDTA, 1 mM dithiothreitol (DTT) and 1% poly vinyl pyrrolidone (PVP) in a chilled mortar and pestle and centrifuge the slurry at 10,000 g for 30 min at 4°C.
2. Collect the supernatant and use it for enzyme assay.

Estimation:

1. Prepare reaction mixture.
2. 1 mL of glutamine followed by 1 mL of 2-oxoglutarate followed by 1 mL of NADPH followed by 200 μL of enzyme extract followed by 1.8 mL of buffer.
3. Do not add 2-oxoglutarate in the blank, instead add 1 mL buffer.
4. Incubate for 15–30 min at 37°C.
5. Record the change in absorbance at 340 nm.
6. The protein content in the extract is determined following Lowry et al.'s (1951) method.
7. Activity is expressed as n mole of NAD (P) H oxidized per minute per milligram protein in enzyme extract.

13.8 GLUTAMINE SYNTHETASE (GS)

Principle: This enzyme has high affinity for ammonia. It catalyzes the following reaction.

$$\alpha\text{-glutamate} + NH_3 + ATP \xrightarrow{Mn^{++}} \alpha\text{-glutamine} + ADP + Pi$$

The activity of the enzyme is measured by estimating the production of inorganic phosphate. GS also catalyzes the γ-glutamyl transfer reaction.

$$\text{Glutamine} + \text{Hydroxylamine} \xrightarrow[\text{Arsenate}]{\text{ADP, } Mn^{++}} \text{Glutamyl hydroxamate}$$

Hence, it can also be assayed by measuring the production of γ-glutamyl hydroxamate. The latter method is described later. The γ-glutamyl hydroxamate is made to react with ferric chloride to produce brown color in acidic medium (Pateman, 1969).

Chemicals required:

- Tris-hydroxymethyl aminomethane
- Hydrochloric acid (HCl)
- α-Glutamine
- Sodium arsenate (disodium hydrogen arsenate)
- Manganese chloride ($MnCl_{2)}$
- Hydroxylamine
- Adenosine diphosphate
- Trichloro-acetic acid
- Ferric chloride
- EDTA
- Dithiothreitol (DTT)
- Glycerol
- Ammonium sulfate $(NH_4)_2SO_4$

Preparation of reagents: Prepare the following reagents in 20-mM Tris-HCl buffer (pH 8.0). The concentration of stock solution is indicated in parentheses.

- α-Glutamine: 0.2 mM (700 mg/12 mL)
- Sodium arsenate: 20 mM (500 mg/10 mL) (disodium hydrogen arsenate)
- $MnCl_2$: 3 mM (83 mg/10 mL)
- Hydroxylamine: 50 mM (278 mg/10 mL)
- Adenosine diphosphate: 1 mM (40 mg/10 mL)
- Ferric chloride reagent: dissolve 10 g trichloro-acetic acid and 8 g ferric chloride in 250 mL of 0.5 N hydrochloric acid.
- Imidazole acetate buffer 50 mM (pH 7.8)
- EDTA 0.5 mM: 1.8612 mg in 100 mL of D.D.H_2O
- Dithiothreitol (DTT) 1 mM
- 20% glycerol: 20 mL in 80 mL of D.D.H_2O

Extraction

1. Extract 1 g plant material in 5 mL of 50 mM imidazole acetate buffer (pH 7.8) containing 0.5 mM EDTA, 1 mM dithiothreitol, 2 mM $MnCl_2$, and 20% glycerol at 4°C.
2. Centrifuge at 10,000 × g for 30 min.
3. If purification is required, precipitate the enzyme with $(NH_4)_2SO_4$ at 60% saturation.
4. Resuspend the precipitate in extraction buffer.
5. Desalt over Sephadex G 25.

Estimation

1. Pipette out the reagents as mentioned later:
   a. 2.0 mL glutamine
   b. 0.5 mL sodium arsenate
   c. 0.3 mL $MnCl_2$
   d. 0.5 mL hydroxylamine

   e. 0.5 mL ADP
   f. 0.2 mL enzyme extract
2. To set a blank, add 2 mL 20 mM Tris-HCl buffer, instead of glutamine.
3. Incubate the reaction mixture for 30 min at 37°C.
4. Stop the reaction by adding 1 mL of ferric chloride reagent.
5. Measure the brown color developed, at 540 nm wavelength in a spectrophotometer.
6. Prepare a range of standards containing 100–500 μg γ-glutamyl hydroxamate in 4-mL buffer solution and develop the color by adding 1 mL of ferric chloride reagent.

13.8.1 CALCULATION

Find out the amount of γ-glutamyl hydroxamate formed in the reaction using the standard graph. Express the enzyme activity as nanomole γ-glutamyl hydroxamate formed per minute per milligram protein.

CHAPTER

Other biochemical traits 14

Crop genotypes also vary in several other biochemical constituents, viz., total phenols, ascorbic acid, alcohol dehydrogenase (ADH) and each one plays specific protective role in plant cells. In this chapter detailed protocol are described.

14.1 TOTAL PHENOLS

Phenols, the aromatic compounds with hydroxyl groups, are widespread in plant kingdom. They occur in all parts of the plants. Phenols are said to offer resistance to diseases and pests in plants. Grains containing high amount of polyphenols are resistant to bird attack. Phenols include an array of compounds such as catechol, caffeic acid, tannins, flavonols, etc. Total phenols estimation can be carried out with the Folin–Ciocalteau reagent.

Principle: Phenols react with oxidizing agent phosphomolybdic acid in Folin–Ciocalteau reagent under alkaline medium and produce blue colored complex (molybdenum blue) which is measured at 650 nm spectrophotometrically (Malik and Singh, 1980).

Chemicals required:

Ethanol (C_2H_5OH)
Folin–Ciocalteau reagent
Sodium carbonate (Na_2CO_3)

Preparation of reagent:

80% Ethanol: 80 mL in 20 mL of D.DH_2O

Extraction:

1. Weigh exactly 1.0 g of the sample and grind it with a pestle and mortar in 10 mL of 80% ethanol.
2. Centrifuge the homogenate at 10,000 rpm for 20 min.
3. Save the supernatant
4. Reextract the residue with five times the volume of 80% ethanol, centrifuge and pool the supernatants.
5. Evaporate the supernatant to dryness.
6. Dissolve the residue in a known volume of distilled water (10 mL).

Phenotyping Crop Plants for Physiological and Biochemical Traits. http://dx.doi.org/10.1016/B978-0-12-804073-7.00014-4

Estimation:

1. Pipette out 0.2 mL of sample into test tubes.
2. Make up the volume in each tube to 3 mL with distilled water.
3. Add 0.5 mL of Folin–Ciocalteau reagent.
4. After 3 min, add 2 mL of 20% sodium carbonate solution to each tube.
5. Mix thoroughly. Place the tubes in a boiling water for exactly 1 min, cool and measure the absorbance at 650 nm against a reagent blank.
6. 100 mg of Catechol in 100 mL water as stock.
7. 10 mL stock makeup the volume to 100 mL with distilled water as working standard.
8. A series of volumes from 0.2 to 1 mL of this standard solution gives a concentration range of 10–100 mg.
9. Then proceed as that of the sample and read the color.

Calculation: From the standard curve find out the concentration of phenols in the test sample and express as mg phenols/100 g material.

In this laboratory, leaf phenol content was validated against the *Aspergillus flavus* infection in groundnut and a negative correlation was reported (Latha et al., 2007).

14.2 ASCORBIC ACID

Ascorbic acid (vitamin C) is present in almost all fresh fruits and vegetables in varying quantities ranging from 0.02 to 1.0 mg per g fr.wt. It is most abundantly found in bitter gourd and berries.

The method for estimation of ascorbic acid is given here as described by Albrecht (1993). It is an easy and simple method based on titration technique.

1. *Titration method*

   Principle: 2, 6-Dicholorophenol indophenol (2, 6-DCPIP) is a blue-colored dye but turns pink when reduced by ascorbic acid. Oxalic acid or metaphosphoric acid may be used as a titrating medium because it increases the stability of ascorbic acid in the medium.

Chemicals required:

- 2,6-Dichlorophenol indophenol (2,6-DCPIP)
- Oxalic acid (or) metaphosphoric acid
- Ascorbic acid standard

Preparation of reagents:

- Standard 2,6-DCPIP solution of concentration 0.5 mg mL^{-1} (50 mg in 100 mL)
- 3% metaphosphoric acid: 3 mL in 100 mL of D.D.H_2O

Extraction:

1. Grind known weight (0.5–5 g as the case may be) of sample using a pestle and mortar with 10–20 mL of 3% Meta phosphoric acid.
2. Centrifuge the macerate at 1000 × g for 10 min.

3. Take the supernatant and make the volume up to 100 mL.
4. Pipette out 5 mL of the supernatant, add 10 mL of 3% metaphosphoric acid, and titrate it against standard 2,6-DCPIP solution of concentration 0.5 mg mL^{-1} until the pink color develops completely, that is, persists for a few seconds.
5. Note down the difference between final and initial volumes of the dye (say V_2 mL).

Estimation: Pipette out 5 mL of the working standard of ascorbic acid (0.1 mg mL^{-1} concentration) in a beaker add 10 mL of 3% metaphosphoric acid and titrate it against the dye. Record the final volume of dye at the end point as mentioned earlier (say V_1 mL).

Calculation: The amount of ascorbic acid in terms of mg/100 g of sample can be calculated as follows:

$$\frac{a}{V_1} \times \frac{V_2(\text{Total volume of sample})}{b(\text{Total weigth of sample taken})} \times 100$$

where

a = 0.5 mg (the concentration of working standard of ascorbic acid = 0.5 mg in 5 mL taken for titration
b = 5 mL, that is, volume of sample taken for titration
V_1 = volume of dye in case of titration with standard solution
V_2 = volume of dye in case of titration with sample solution

2. *Colorimetric method*
Ascorbic acid can also be determined colorimetrically. Ascorbic acid is first dehydrogenated by bromine and then treated with 2,4-dinitroplenylhydrazine (DNPH) to form osazone. Osazone, when dissolved in sulfuric acid, gives orange-red-colored solution. OD is measured at 540-nm wavelength using a colorimeter.

Chemicals required:
- 2,4-dinitrophenylhydrazine (DNPH)
- Sulfuric acid (H_2SO_4)
- Bromin water
- Oxalic acid
- Thiourea
- Ascorbic acid standard

Preparation of reagents:
- 2% DNPH reagent is prepared by dissolving 2 g of the chemical in 0.5N H_2SO_4 and making the volume up to 100 mL.
- Bromine water is prepared by just dissolving one to two drops of liquid bromine in 100 mL of distilled water.
- 4% oxalic acid solution: 4 g in 100 mL D.DH_2O.
- 10% Thiourea solution: 10 g in 100 mL of D.DH_2O.
- 80% sulfuric acid: 20 mL of sulfuric acid makeup to 100 mL.

Extraction:
The process of extraction of ascorbic acid from the sample is the same as in case of "Titration Method." Preserve the supernatant and follow the following steps:

Estimation:
1. Take 10 mL of the aliquot in a conical flask.
2. Add bromine water (dropwise) till the solution turns orange-yellow in color.
3. Expel excess of bromine by blowing-in air.
4. Add 4% oxalic acid and make up the volume up to 25 mL.
5. Take 10 mL of stock ascorbic acid solution and brominate it the same way as above (as the sample).
6. Pipette out 10–100 μL standard brominated ascorbic acid into a series of test tubes.
7. Pipette out 0.1–2.0 mL of brominated sample extract into another series of test tubes.
8. Add distilled water in each test tube and make up the volume up to 3 mL.
9. Add 1 mL of 2% dinitrophenylhydrazine reagent and thereafter one to two drops of 10% thiourea solution to each test tube. Shake the tubes thoroughly and incubate them at 37°C for 2–3 h.
10. Add 10 mL of 80% sulfuric acid into each test tube, so that the orange-red osazone crystals get completely dissolved.
11. Measure OD at 540 nm by a colorimeter/spectrophotometer. Set a blank as above but with water in place of ascorbic acid solution.

Calculation: Ascorbic acid content in the sample can be calculated by plotting a graph showing ascorbic acid concentration on "X" axis and their respective OD values on "Y" axis.

14.3 ALCOHOL DEHYDROGENASE (ADH)

Alcohol dehydrogenase (ADH) catalyzes the anaerobic oxidation of acetaldehyde, a product of pyruvate oxidation, to ethanol. The enzyme is important, as it utilizes NADH, and thus allows the glucose metabolism by glycolysis. Measurement of this enzyme in crop plants exposed to submerged conditions helps in understanding the level of susceptibility or tolerance by genotypes.

Principle: The assay described by Chung and Ferl (1999) utilizes the reverse reaction, that is, oxidations of ethanol by ADH with the help of NAD, resulting in the synthesis of acetaldehyde and NADH. The increase in absorbance due to NADH at 340 nm is estimated spectrophotometrically.

Required chemicals
- Tris HCl
- Dithiothreitol (DTT)
- Hydrochloric acid

- Nicitinamide adenine dinucleotide (NAD)
- Ethanol

Preparation of reagents

Extraction buffer: (Tris HCl 50 mM + 15 mM DTT, pH 8)

Extraction buffer is prepared by dissolving 0.606 g Tris (hydroxyl methyl) aminomethane, 0.231 g DTT in 50 mL D.D.H_2O. pH is adjusted with the help of a pH meter by using 0.1N HCl. Final volume is made to 100 mL to get a solution of desired molarity.

- *Tris buffer (150 mM, pH 9.0)*: Tris buffer is prepared by dissolving 3.633 g of Tris in distilled water and volume is made to 100 mL with distilled water. pH is adjusted with the help of a pH meter by using 0.1 N HCl. Final volume is made to 200 mL to get buffer of desired molarity.
- *NAD (13.005 mM)*: NAD is dissolved in distilled water and final volume is made to 25 mL in a volumetric flask.
- *Ethanol (60% v/v)*: Absolute alcohol chemical grade, 60 mL is diluted to 100 mL in a volumetric flask.
- All solutions are preserved at 4°C.

Extraction:

1. Plant tissue of 0.5 g is first pulverized with liquid nitrogen and then homogenized with 5.0 mL of extraction buffer.
2. The extract is centrifuged at 12,000 rpm for 15 min at 4°C in a refrigerated centrifuge, and the supernatant is used as a source of enzyme.

Estimation:

1. Three milliliters of reaction mixture contains 100 μL of enzyme extract, 1 mL of 150 mM Tris buffer, 0.2 mL of NAD, and 1 mL of 60% ethanol.
2. Finally make up to 3 mL with water.
3. Reaction mixture except NAD is prepared in test tubes, and each sample can be used as blank to adjust zero.
4. NAD is added to initiate the reaction and increase in absorbance at 340 nm is recorded for 1 min.

Calculation: Amount of NADH formed is computed by drawing a standard curve of NADH at 340 nm, and activity is expressed as nmol. NADH formed per milligram protein per minute.

14.4 GLYCINE BETAINE

Glycine betaine belongs to group of compounds commonly known as quaternary ammonium compounds. It is a derivative of glycine. It is reported to accumulate in many plant species under drought, salinity, and temperature (high and low) stresses. Its precursor is choline and two enzymes, viz., choline monooxygenase and betaine aldehyde dehydrogenase play crucial role in its synthesis in bacteria and plants.

It serves as an osmolyte by lowering the osmotic potential of the cell and thus prevents movement of water from the cell, as well as compatible solutes by preventing denaturation of macromolecules such as enzymes/proteins. Glycine–betaine estimation is done in dried leaf powder as per the method of Greive and Grattan (1983).

Principle: The assay is based on the fact that at low temperature betaine makes a betaine–periodite complex with iodide in acidic medium, which absorbs at 360 nm in UV range.

Chemicals required

- Potassium iodide
- Iodine
- Sulfuric acid

Preparation of reagents

Cold potassium iodide–iodine solution: Iodine (15.7 g) and potassium iodide (20 g) were dissolved in 100 mL of water and kept in refrigerator at 4°C.
Sulfuric acid (2 N): Fifty-five milliliters of sulfuric acid is dissolved in distilled water and the volume made up to 1 L.

Extraction

1. Extract prepared by finely ground dry plant material (0.5 g) is mechanically shaken with 20 mL of deionized water for 48 h at 25°C.
2. The samples are then filtered and the filtrate is stored in freezer until analysis.
3. Thawed extracts are diluted 1:1 with 2 N sulfuric acid.
4. Aliquot (0.5 mL) is measured into test tube and cooled in ice water for 1 h.
5. Add cold potassium iodide–iodine reagent (0.2 mL) and gently mix with vortex mixture.
6. Store the samples at 0–4°C for 16 h.
7. After the expiring of the period samples are transferred to centrifuge tubes and then centrifuged at 10,000 g for 15 min at 0°C.

Estimation

1. The supernatant is carefully aspirated with 1 mL micropipette.
2. As the solubility of the periodite complexes in the acid reaction mixture increases markedly with temperature, it is important that the tubes be kept cold until the periodite complex is separated from acid media.
3. The periodite crystals are dissolved in 9 mL of 1,2-dichloro ethane (reagent grade).
4. Vigorous vortex mixing is done to effect complete solubility in developing solvent.
5. After 2.0–2.5 h the absorbance is measured at 365 nm with UV–VIS spectrophotometer.
6. Reference standards of glycine–betaine (50–200 μg/mL) are prepared in 2 N sulfuric acid and the procedure for sample estimation was followed.

CHAPTER

Plant pigments

15

15.1 CHLOROPHYLLS

1. The chlorophylls are the essential components for photosynthesis, and occur in chloroplasts as green pigments in all photosynthetic plant tissues.
2. They are bound loosely to proteins but are readily extracted in organic solvents such as dimethyl sulfoxide (DMSO), acetone, or ether.
3. Chemically, each chlorophyll molecule contains a porphyrin (tetrapyrrole) nucleus with a chelated magnesium atom at the center and a long-chain hydrocarbon (phytol) side chain attached through a carboxylic acid group.
4. These pigments are located in the chloroplasts of the plant.
5. The energy of sunlight is captured by chlorophyll pigments to make food during the process of photosynthesis.
6. Equation for photosynthesis in a simple form would be:
   Water + Nutrients in soil + Carbon dioxide + sunlight → food for plants + Oxygen
7. There are at least five types of chlorophylls in plants. Chlorophylls a and b occur in higher plants, ferns, and mosses. Chlorophylls c, d, and e are only found in algae and certain bacteria.

15.1.1 ESTIMATION OF CHLOROPHYLL

a. By Acetone Method

Chlorophyll is soluble in acetone. When the sample is macerated in acetone, chlorophyll gets dissolved in it. The optical density of the extract is measured at 663 and 645 nm wavelengths using spectrophotometer because at these wavelengths, maximum absorption of chlorophyll "a" and "b" takes place respectively.

Principle: Chlorophyll is extracted in 80% acetone and the absorbances are read at 663 and 645 nm in a spectrophotometer. Using the absorption coefficients, the amount of chlorophyll is calculated (Arnon, 1949).

Chemical required:

Acetone

Preparation of reagent:

80% acetone: 80 mL in 20 mL of D.DH_2O

Phenotyping Crop Plants for Physiological and Biochemical Traits. http://dx.doi.org/10.1016/B978-0-12-804073-7.00015-6

Extraction

Place 0.2 g of fresh plant leaf material in a test tube and grind with 10 mL of 80% acetone in mortar using pestle.

Estimation

1. Chlorophyll extracted into acetone solution was collected to from test tubes by filtering the homogenate using what man filter paper no. 1.
2. Wash out the homogenate two to three times with 5 mL of 80% acetone each time.
3. Make the final volume of the filtrate up to 25 mL with 80% acetone.
4. Concentration of chlorophylls a, b and total chlorophyll were quantified in samples by reading the optical density at 663 and 645 nm.
5. Calculate the chlorophylls a, b and total chlorophyll using the formula.

Calculation: Calculate the amount of chlorophyll present in the extract in mg chlorophyll per g tissue using the following equations:

$$\text{mg chlorophyll a/g tissue} = 12.7(A663) - 2.69(A645) \times \frac{V}{1000 \times W}$$

$$\text{mg chlorophyll b/g tissue} = 22.9(A645) - 4.68(A663) \times \frac{V}{1000 \times W}$$

$$\text{mg total chlorophyll/g tissue} = 20.2(A645) + 8.02(A663) \times \frac{V}{1000 \times W}$$

where

A = absorbance at specific wavelengths
V = final volume of chlorophyll extract
W = fresh weigh of tissue extracted

b. By DMSO (Dimethyl Sulfoxide) Method

Principle: Chlorophyll is extracted in 80% acetone or DMSO and the absorbances are read at 663 and 645 nm in a spectrophotometer. Using the absorption coefficients, the amount of chlorophyll is calculated (Hiscox and Israelstam, 1979).

Chemical required:

DMSO (dimethyl sulfoxide)

Extraction:

1. Place 0.1 g of fresh plant leaf material in a 100-mL volumetric flask and add 10 mL of DMSO.
2. Keep the conical flasks for overnight.

Estimation:

1. Chlorophyll extracted into DMSO solution was collected from conical flasks and concentration of chlorophylls a, b and total chlorophyll were quantified by reading the optical density at 663 and 645 nm.
2. Calculate the chlorophylls a, b and total chlorophyll using the formula.

Calculation: Calculate the amount of chlorophyll present in the extract in mg chlorophyll per g tissue using the above equations.

15.2 CAROTENOIDS

- They are generally red, orange, or yellow pigments. Familiar carotene gives a carrot their color.
- They do not mix completely in water.
- They pass the absorbed sunlight energy to chlorophyll, and so assist in the process of photosynthesis.
- The pigment found in carrot is beta carotene.
- Lycopene gives a red color to a ripe tomato or to a red bell pepper.
- Other fruits and veggies where these pigments are found are mango, melon, apricot, sweet potatoes, parsley, and spinach.
- There are two kinds of carotenoids:
  - Carotenes
  - Xanthophylls

15.2.1 QUANTIFICATION OF CAROTENOIDS IN GREEN LEAVES

a. By HPLC Method

In plants and eyes, carotenoids play an important role in absorption of blue light and exert antioxidant activity preventing damages caused by light and free radicals. As leafy vegetables are widely available and easy to gather from the wild or in agro-ecosystems, or may be cultivated at low cost, their consumption and conservation is being promoted for increased health benefits (Johns, 2007). Quantification of carotenoids was standardized by the HPLC method using UV detector in this laboratory (Julie et al., 2010) and the protocol is given below:

Chemicals required:

- HPLC acetone
- HPLC diethyl ether
- HPLC petroleum ether
- HPLC acetonitrile
- HPLC ethyl acetate
- HPLC methanol
- HPLC triethyl amine
- Glass wool
- Sodium sulfate
- Sodium chloride
- Celite
- Nitrogen gas

Extraction:

1. Select a clean sample.
2. Cut and mix in a blender.
3. Weigh 3 g of sample and deposit in a glass covered with foil.
4. The sample should not be exposed to light.
5. Transfer the weighed sample into the pestle and mortar for grinding.

6. It should be grind with acetone with enough celite for aggressive grinding and for easy filtration until entire green mater comes out from the mater.
7. After grinding filter the sample through vacuum filtration.
8. Meanwhile add sufficient acetone for complete extraction from the mixer.
9. Sample should be collected into 500-mL jar.

Partition to ethyl ether and petroleum ether:

1. Pour 25-mL petroleum ether and 25-mL di ethyl ether in a separatory funnel (It should be properly cleaned without any moisture inside).
2. Add small portion of the acetone extract to the petroleum ether–ethyl ether mixture.
3. Then add nearly about 300–400 mL distilled water in a elevated portion not in a straight position.
4. Then tight the funnel to the stand and allow it for few minutes.
5. Cover the funnel with aluminum foil during entire partition.
6. After few minutes open the separatory funnel to allow the water to be collected into the flask.
7. This is said to be the "first wash."
8. Then add remaining half of the amount of acetone extract into the funnel.
9. Again add 300–400 mL distilled water.
10. Keep it for a few minutes by covering with aluminum foil.
11. Then again remove the water. This is called second wash. Repeat the washing up to five times with distilled water to remove residual acetone.
12. If an emulsion is formed, add one to two teaspoons of sodium chloride to the distilled water, mix, and wash the extract.
13. No water will be present in the sample after five washings.

Method of filtration:

1. After complete five times washing, the sample is filtered through the funnel into the 250-mL round bottom flask (RBF) covered with aluminum foil.
2. Place sufficient glass wool into funnel and put sodium sulfate above the glass wool.
3. The sodium sulfate should be completely closed by the glass wool for proper filtration.
4. Filter the sample from the separatory funnel to the round bottom flask via sodium sulfate.
5. After complete filtration, the sample should be transferred for evaporation.
6. Add diethyl ether for complete green matter filtration.

Evaporation of the sample:

1. Place the round bottom flask to the neck of rotary evaporation with the help of clamp and run the evaporation process until 2–3 mL of sample will remain in the RBF.
2. RBF should be closed with some cloth against light or aluminum foil.

Evaporation with nitrogen gas:

1. Take the RBF from rotary evaporation and place it into the container with complete aluminum foil (The sample does not expose to light).

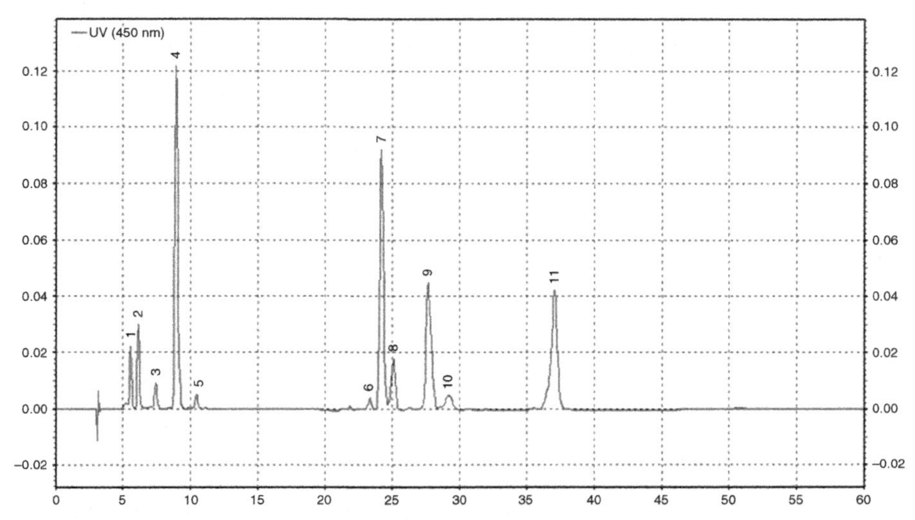

FIGURE 15.1 Chromatogram Showing Carotenoid Standard Peaks: 1- Neoxanthin; 2- Violaxanthin; 4- Lutein; 7- Chlorophyll B; 9- Chlorophyll A; 11- beta-Carotene.

2. Then evaporate the remained sample having petroleum ether and diethyl ether with nitrogen gas.
3. Keep the sample at –20°C and tightly hold the flask with par film.

HPLC instrument set-up protocol for estimation of carotenoids:

1. Instrument set-up protocol for estimation of carotenoids through HPLC in green leafy vegetables was standardized. The chromatogram showing carotenoids standard peaks is shown in Fig. 15.1.
2. Standardization was done by changing the mobile phase concentrations, flow rates, and injection volume of the standard to be injected.
3. Finally, chromatogram with sharp peaks at particular retention time was obtained.
4. Take the sample from the freezer (−20°C), allow it to 2 min, and then dilute the sample with 10 mL of HPLC acetone (dilution depends on the intensity of color of the solution).
5. If the sample is thick, dilute more than 10 mL (i.e., 15 mL).
6. Then pipette out the diluted sample from the RBF and then filter through a 0.2-μm syringe filter into the HPLC vial.
7. Place the HPLC vial in autosampler for analysis.
8. The instrument set-up protocol for carotenoid estimation is given below.

Mobile phase:

Pump A: HPLC acetonitrile with 0.05% HPLC triethyl amine (TEA)
Pump B: HPLC methanol: HPLC ethyl acetate in 1:1 ratio

- Flow rate: 1.0 mL/min
- Column symmetry: C-18

- Injection volume: 10 μL
- Detection: UV detector at 450 nm
- Gradient: binary
- Run time: 60 min

b. By UV–VIS Spectrophotometer Method

Principle: The sample extract, obtained for the estimation of chlorophylls (please refer to Acetone Method), can also be used for the quantification of carotenoids. The absorbances are read at 663 nm, 645 nm, and also at 480 nm in a spectrophotometer. Using the absorption coefficients, the amount of carotenoids is calculated.

Extraction: The sample extract, obtained for the estimation of chlorophylls (please refer to Acetone Method), can also be used for the quantification of carotenoids.

Estimation: In addition to measurement of absorbance of the extract at 663 and 645 nm, spectro photometric readings are also recorded at 480-nm wavelength.

Calculation: Carotenoid content is calculated using the formula (Price and Hendry, 1991) given here:

$$\text{Total carotenoids (mg/g fw)} = [A_{480} + (0.114 \times A663) - (0.638 - A645)] \times \frac{V}{1000} \times W$$

This equation compensates for the interference at this wavelength, due to chlorophyll.

15.3 LYCOPENE

Lycopene is having a molecular formula of $C_{40}H_{56}$ and is responsible for the red color of tomato and watermelon. Lycopene attracts the consumer and enhances the marketability of these fruits although it has no nutritional value.

Principle: The lycopene in the sample is extracted in acetone and then taken up in *n*-hexane. Lycopene has absorption maxima at 473 and 503 nm.

Chemicals required:

- Acetone 100%
- *n*-hexane

Extraction:

1. Take 1.0 g of sample (pericap of ripe tomato), grind it in prechilled pestle and mortar with 10 mL of acetone (100%). Centrifuge at 10,000 × g for 5 min.
2. At low temperatures (4°C). Collect the supernatant and preserve it.
3. Extract the residue (pellets) with acetone (three to four times) by centrifugation and collect the supernatant after each centrifugation until the residue (pellets) becomes colorless.

4. Transfer all the supernatant in a volumetric flask and measure the total volume.
5. While grinding there should be minimum light. Maintain low temperature.

Estimation:

1. Take 10 mL of the supernatant and add equal volume of *n*-hexane and the bottom layer is of acetone.
2. The bottom layer is discarded. Measure OD of the upper layer (separated as above) at 503-nm wavelength using a spectrophotometer or colorimeter.
3. Blank is prepared by mixing equal volumes of acetone and *n*-hexane.
4. In a cuvette, this mixture is taken along with 0.5 mL of distilled water.
5. This serves as blank. It may be mentioned that one of the peak absorption values of lycopene is also at 473 nm; however, this wavelength is very close to the absorption maximum of carotenoids (480 nm).
6. Therefore, for quantitative estimation of lycopene, the OD value at 503 nm only is taken into consideration (Ranganna, 1976).

Calculation:

$$\text{Lycopene}\,(\mu g/g\,\text{fr.wt.}) = \frac{3.121 \times \text{OD value at } 503\,\text{nm} \times \text{vol. of sample} \times \text{Dilution factor}}{\text{Fresh weight of sample (g)}}$$

15.4 ANTHOCYANIN

Anthocyanins constitute an important group of plant pigments. They are water soluble and belong to the family of flavonoids. More than 500 different anthocyanins have been identified. These pigments give plants, flowers, and fruits, their brilliant colors ranging from pink through scarlet, purple, and blue.

Anthocyanins are particularly found in fruits such as grapes, purple grapes, black berries, strawberries, and raspberries.
In addition to being powerful antioxidants, anthocyanins also possess antiinflammatory, antimicrobial, and anticancer properties.
Anthocyanin content in flower petals is generally measured by the method of Ronchi et al. (1997).

Chemicals required:

- Hydrochloric acid
- Ethyl alcohol

Reagent preparation:

- 1% Hydrochloric acid

Extraction: Take 100 mg of petals. Extract with 10 mL of 1% HCL in ethyl alcohol. Centrifuge twice at 10,000 rpm for 5–10 min at room temperature and collect the supernatants.

Estimation: Measure OD at 530-nm wavelength in a spectro-photometer.
Calculation: Express anthocyanin content as A530 per gram fresh weight.

CHAPTER

Growth regulators

16

Plant hormones and growth regulators are compounds that influence flowering, aging, root growth, prevention or promotion of stem elongation, prevention of leaf emergence, and/or leaf fall and many other physiological processes. These substances produce major growth changes at a very low concentration.

Hormones are endogenously produced substances in plants, while plant growth regulators may be synthetic compounds (ie, IBA and Cycocel) that mimic naturally occurring plant hormones or they may be natural hormones that are extracted from plant tissue (eg, IAA). These growth regulating substances, most often, are applied as a spray to foliage or as a liquid drench to soil around a plant's base. Generally, their effects are short-lived and they may need to be reapplied to achieve the desired effect.

Plant growth regulating substances are normally classified into five groups, viz.: (1) Auxins, (2) Gibberellins, (3) Cvtokinins, (4) Ethylene, and (5) Abscisic acid. For the most part, each group contains both naturally occurring hormones and synthetic substances. In this chapter, the methods of quantitative estimation of ABA, IAA, GA, and ethylene in plant samples are described.

16.1 ESTIMATION OF INDOLE ACETIC ACID (IAA)

Spectrofluorimetric method is followed for the estimation of indole acetic acid in plant tissues as described by Knegt and Bruinsma (1973).

Chemicals required:

- Methyl alcohol
- Potassium monohydrogen phosphate (K_2HPO_4) (0.5 M)
- Petroleum ether
- Diethyl ether
- Phosphoric acid (2.8 M)
- Trifluroacetic acid–acetic anhydride reagent

Preparation of reagents:

- Potassium monohydrogen phosphate (K_2HPO_4) (0.5 M): 8.709 g in 100 mL of D.D.H_2O.
- Phosphoric acid (2.8 M)

Extraction:

1. Grind 2 g of plant sample in 10-mL methyl alcohol using a pestle and mortar.
2. Filter the homogenate through G4 glass filter under vacuum condition.

Phenotyping Crop Plants for Physiological and Biochemical Traits. http://dx.doi.org/10.1016/B978-0-12-804073-7.00016-8

3. Extract the residue thrice using 10-mL methanol each time and collect the filtrate.
4. Evaporate the filtrate at 30°C (using a rotary evaporator) to get an aqueous residue.
5. Pour the aqueous residue into a beaker and adjust its pH to 8.5 with K_2HPO_4 solution (0.5 M).
6. Take the content of beaker into a separating funnel.
7. Extract it thrice; two times each with 10 mL each of petroleum ether and the third time with 10 mL of diethyl ether (at each step of the extraction, upper lipid fraction in the separating funnel should be discarded and lower aqueous layer should be used).
8. Collect this aqueous layer, which contains IAA, into a beaker and adjust its pH to 3.0 by adding phosphoric acid solution (2.8 M).
9. Extract IAA finally with 10 mL of diethyl ether.
10. Evaporate the ether under vacuum condition and preserve the residue.
11. Dissolve the residue in a 5–10 mL of cold methanol.

Estimation:

1. One milliliter of the methanolic extract of IAA is added into each of the six test tubes (kept in a test tube stand).
2. To each test tube, add 1 mL of methyl alcohol containing 0, 5, 10, 20, 30, or 40 ng of IAA, respectively.
3. Dry the content under vacuum condition.
4. Add 0.2 mL of ice cold trifluroacetic acid–acetic anhydride reagent and mix well.
5. Incubate the tubes on ice for at least 15 min.
6. Add 3-mL water in each tube to stop reaction.
7. A blank may be prepared by adding 3-mL water to one of the six tubes and 0.2-mL reagent after 15 min of incubation (as discussed earlier). Measure OD in a spectrofluorimeter. (Excitation wavelength is kept at 440 nm and emission at 490 nm).
8. Calculate the amount using a standard curve. To prepare trifluoroacetic acid–acetic anhydride reagent, add equal volume of each chemical and mix thoroughly.
9. Both chemicals must be precooled at 0°C. Reagent must be ice-cold before use. Incubation should also be carried out under ice-cold condition.

16.2 ESTIMATION OF GIBBERELLINS

For quantitative estimation of gibberellins, gas liquid chromate-graphy (GLC) is used as described by Shindy and Smith (1975).

Chemicals required:

- Methyl alcohol
- Hydrochloric acid
- Ethyl acetate

Preparation of reagents:

- Methanol (80%)
- 1 % Hydrochloric acid

Extraction:

1. Grind 5 g of fresh sample in 80% (v/v) cold aqueous methanol with a mortar and pestle.
2. The macerate is transferred to a 100-mL flask and the volume is adjusted to 20 mL of methanol for each gram fresh weight of sample.
3. The tissue is allowed to extract for 24 h at 0–4°C and then is vacuum filtered through Whatman No. 42 filter paper.
4. The residue along with filter paper is returned to the flask with a fresh volume of methanol, shaken for 30 min on a shaker, and filtered again.
5. The procedure is repeated once more and the combined extract is evaporated to the aqueous phase in a rotary flask evaporator.
6. Take 15–30 mL of the aqueous phase, adjust pH to 2.8 with 1% HC1, and partition three times with equal volume of ethyl acetate.
7. The combined acidic ethyl acetate fraction is reduced in a volume to be used for GLC determination of GA. Take 0.5– 0.1 mL of the above sample in replicates.
8. Place in 1-mL reactive vials. Dissolve in 100 μL of N, O-bis (trimethyl-silyl) acetamide to prepare trimethylsilyl derivative (silylation).
9. The vials are immediately capped and heated over a hot plate (50°C) for 30 min before GLC analysis.

Estimation:

Gas Liquid Chromatography

1. 1–3 μL of the trimethylsilyl derivative is injected into a GLC equipped with a flame ionization detector, a temperature programmer, and a glass column SE-30 (2 m long having 2-mm diameter).
2. The GLC is operated under the following conditions:
   Hydrogen flow: 60 mL min^{-1}
   Injector temperature: 200°C
   Detector temperature: 275°C
   (Temperature programming begins at 100°C and is increased linearly to 250°C at the rate of 10°C min^{-1})
3. The retention time and temperature for each peak are recorded and compared to that of trimethylsilyl derivative of authentic standard.

Calculation: The total area at retention time is used for calculating the amount of GA. The amount is calculated with the help of standard peak and expressed on ng per g fresh weight basis.

16.3 ESTIMATION OF ABSCISIC ACID (ABA)

The method of abscisic acid estimation in leaves, as given by Zeevaart (1980), is described here with a little modifications or amendments using HPLC.

Sample collection: Fully expanded leaves should be collected in butter paper bags and kept in an ice box. The samples are preserved in liquid nitrogen for further analysis.

Chemicals required:

- Acetic acid
- Ethyl alcohol

Preparation of reagents:

- 1% Acetic acid

Extraction:

1. Take leaf sample (2–5 g) in a flask and pour so much volume of acetone containing 1% acetic acid that the sample dips into it completely and incubate it overnight at 4°C.
2. Filter the extract through a filter paper (Whatman No. 4).
3. The extraction should be repeated twice or thrice (ie, repeat the above steps two to three times; each time use the residue on the filter paper for extraction).
4. Evaporate acetone from the total volume of extract using a rotary evaporator at 40–50°C till some residue is left on the bottom.
5. Preserve the residue in a cool and dark place.

Sample preparation:

1. Take out the flask with the residue (sample) as mentioned earlier.
2. To it add 2-mL distilled water having 1% acetic acid (v/v) and sonic ate thoroughly.
3. Transfer the content of the flask into sample vials.
4. Make the volume in each sample vial up to 5 mL with distilled water containing 1% acetic acid (v/v). Store the sample vials for assay using HPLC.
5. Generally, the experimental conditions, for the analysis of ABA using HPLC, are maintained as follows:
6. Column: reverse phase C_{18} column having particle size equal to 5 μrn.
   Detector: UV detector
   Wavelength: 265 nm
   Flow rate: 1.5 mL min^{-1}
   Calibration curve: using 10-ppm solution of ABA in 95% ethyl alcohol.

Solvents:

A: 1% acetic acid in distilled water (HPLC grade) v/v.
B: 1% acetic acid in methyl alcohol (HPLC grade) v/v.

Both the solvents are filtered through Millipore nylon filter (13 μM) with the help of vacuum pump solvent filter kit.

Estimation:

1. Filter the standard solution of ABA through nylon filter (0.45-nm pore size) and inject 10 μL of this solution into HPLC.
2. Record the peak which generally appears after 30–50 s.
3. Inject 10 μL of the sample into the HPLC and record the peak area.

Calculation: The ABA content in the sample may be calculated using the equation given here:

$$\text{in } 10\,\mu\text{L solution (ng) ABA content (ng)}$$
$$= \frac{\text{Peak area}\left(\text{cm}^2\right)\text{of sample} \times \text{Amount of ABA}}{\text{Peak area}\left(\text{cm}^2\right)\text{of standard solution} \times \text{Fresh weigth of the leaf sample (g)}}$$

16.4 ESTIMATION OF ETHYLENE

Ethylene, a ripening hormone, is present in very small quantity in plants. It is estimated using a gas chromatograph. One may find the details regarding the estimation procedure in the instrument manual; however, for the benefit of students at postgraduate level, the basic of chromatography along with the ideal operating conditions and sample collection method are given here. It may be mentioned that in plant physiological studies, ethylene is generally assayed during different stages of ripening of climacteric fruits such as mango (Reporter, 1987).

Sample Collection

1. Put mango fruits (known weight) in a cylinder or jar.
2. Seal the mouth with a gasket.
3. Incubate for at least 2–4 h at room temperature.
4. Withdraw the gas sample (ethylene) using a hypodermic syringe for assay. A known volume of gas should be assayed.

Ideal operating conditions for gas chromatograph

Carrier gas: hydrogen/nitrogen mixture
Flow rate: 20–30 mL min^{-1}
Gas for detector: hydrogen and air
Column: Porapak-Q 80/100 mesh packed
Oven/column temperature: 60°C
Injector temperature: 110°C
Detector temperature: 85°C
Detector (to be used): flame ionization detector (FID)
Retention time for ethylene: 1.3 min

1. The carrier gas from a cylinder is passed through flow regulator to an injection port, where it picks up the sample for analysis.
2. The carrier gas + sample mixture then passes through the column in a thermostatic oven where the components of the mixture are separated.
3. The area of the peak depends upon the amount of substance present, the detector efficiency, and the degree of amplification used.
4. If the latter factor is hold constant, the recorded peak area is a direct measure of the amount of substance present in the sample.
5. Prior to assay of sample, the instrument must be calibrated with a known volume of standard ethylene gas.
6. For calculating the amount of ethylene produced per gram of sample per unit time, refer to "Nitrogene" estimation.

SECTION V

CHAPTER

Analytical techniques

17

17.1 ULTRAVIOLET VISIBLE (UV–VIS) SPECTROPHOTOMETER

Introduction: According to electromagnetic theory of light, light travels in the form of waves described by three attributes – wavelength, frequency, and energy. The visible portion of electromagnetic spectrum extends from 360 to 900 nm and the ultraviolet (UV) from 200 to 380 nm. The spectrophotometer uses a tungsten lamp for measurements in the visible region (360–900 nm) and a deuterium lamp for the ultraviolet region (200–380 nm) of the spectrum. A monochromator consisting of a grating or a prism or a combination of both is used to obtain a narrow band of wavelengths continuously selectable through the exit slit of the monochromator. The beam of light transmitted by the sample is detected with the help of a suitable detector either a phototube or a photomultiplier tube (for greater sensitivity) and the optical density displayed on a suitable analogue or a digital read out.

Light that includes all the rays in the wavelength range of visible rays is called white light. When white light is irradiated on some substance and the substance absorbs the blue light, it appears yellow, which is the (additive) complementary color of blue. If blue monochromatic light is irradiated on this substance, the light is absorbed and the substance appears black, indicating that no color exists (Table 17.1).

Principle: Spectrophotometer measures the light absorbed by a sample solution at a given wavelength. The absorption of light by molecules is governed by Lambert–Beer law. When light with the intensity of I_0 is irradiated on a certain substance and the light with the intensity of I has transmitted, the following relational formula is established, where K stands for proportional constant.

At this time, I/I_0 is called transmittance (T), $(I/I_0) \times 100$ is percent transmittance (%T) and $(I/T) = \log (I_0/I)$ is called absorbance (Abs). The absorbance (optical density) of a solution is directly proportional to the concentration of the solute (Beer's law) and optical path length (Lambert's law) through the solution.

Objective: Quantitative estimation of chemical substances was measured through ultraviolet visible (UV–VIS) spectrophotometer (Fig. 17.1). The spectropho-tometer consists of three different modes, viz., photometric, spectrum, and kinetic modes. Spectro photometer measures the optical density values of different chemical substances (proteins, carbohydrates, amino acids, chlorophylls, etc.) of known wavelength through photometric mode. Spectrum mode scans the chemical substances of

Phenotyping Crop Plants for Physiological and Biochemical Traits. http://dx.doi.org/10.1016/B978-0-12-804073-7.00017-X

Table 17.1 Relation Between Wavelength, Color, and Its Complementary Color

Sr. No	Wavelength (nm)	Color	Complementary Color
1	400–435	Violet	Yellow green
2	435–480	Blue	Yellow
3	480–490	Greenish blue	Orange
4	490–500	Bluish green	Red
5	500–560	Green	Purplish red
6	560–580	Yellow green	Violet
7	580–595	Yellow	Blue
8	595–610	Orange	Greenish blue
9	610–680	Red	Bluish green
10	680–700	Purplish red	Green

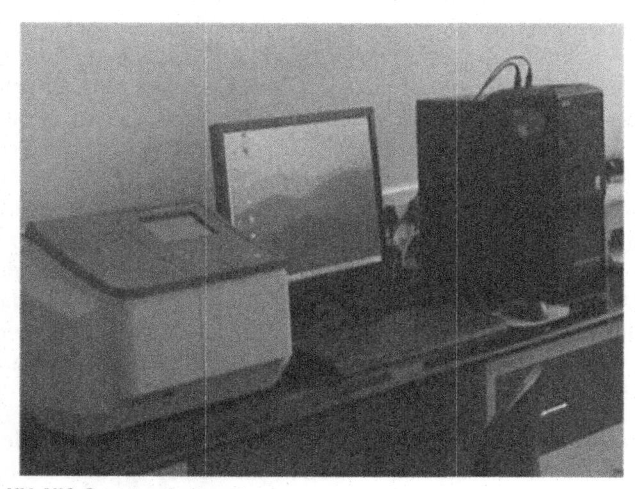

FIGURE 17.1 UV–VIS Spectrophotometer.

unknown wavelength (scanning nano size chemical substances), whereas the kinetic mode of spectrophotometer measures the enzyme activity of a substance (antioxidant enzymes, nitrate reductase, nitrogenase, etc.).

17.2 THIN LAYER CHROMATOGRAPHY (TLC)

Chromatography is highly useful in research laboratories to separate, identify, and characterize unknown compounds.

Principle: Separation of closely related compounds in a mixture through equilibrium distribution of the components between two immiscible phases, viz., the stationery phase and the mobile phase.

TLC is an easy technique for separation and identification of unknown compounds and requires simple apparatus. They readily provide qualitative information and it is also possible to obtain quantitative data with careful attention. A variety of small molecules such as amino acids, sugars, organic acids, and lipids are separated by thin layer chromatography (TLC).

Chemicals required:

silica gel for TLC
butanol
acetic acid
ninhydrin as spraying agent

Preparation of solvent system: Solvent system is prepared using butanol, acetic acid, and distilled water in the ratio of 50:10:40 mL. Pour about 100 mL of the solvent system into the glass tank to a depth of 5 cm. Allow it to stand for 1 h with a lid over the top of the tank.

Preparation of sample:

Sample 1: 50 mg of tryptophan was dissolved in 1 mL of water.
Sample 2: 50 mg of methionine was dissolved in 1 mL of water.
Sample 3: 50 mg of proline was dissolved in 1 mL of water.

Preparation of spraying reagent (w/v): 2% ninhydrin in acetone was prepared and used as a spraying agent.

Preparation of silica gel and silica gel plates:

1. About 40 g of silica gel is dissolved in 100 mL of distilled water and is mixed thoroughly to become slurry.
2. Glass plates are cleaned with ethanol and coat the glass plates with the above silica gel slurry uniformly with the help of a spreader.
3. The plates are dried at room temperature at overnight and then heat the plates in an oven for 1 h.

Procedure:

1. Leave 2 cm from one end of the glass plate and apply the sample containing amino acid with the help of micropipette as small spots.
2. Allow the sample spots to dry and place the glass plates inside the solvent chamber.
3. The solvent rises by capillary action.
4. The chamber was covered with a lid and grease was applied to prevent the entry of air into the chamber until the solvent reaches top of the glass plate.
5. Remove the glass plates from the tank, dry and proceed for the identification of the separated compounds.
6. Spray with ninhydrin reagent for identification of amino acids.

Calculation:

The RF value of each sample was calculated as

$$RF = \frac{\text{Distance traveled by solute}}{\text{Distance traveled by solvent}}$$

17.3 GAS CHROMATOGRAPHY (GC)

17.3.1 INTRODUCTION

Gas chromatography involves separation and analyses of different constituents of mixtures by a mobile gas phase passing over a stationary adsorbent. The technique is similar to column chromatography, except that the mobile phase is replaced by a moving gas that is called the carrier gas.

Gas chromatography is a powerful tool for the analyses of organic materials. It is very handy for low levels of pesticides and other contaminants of the environment. Gas chromatography can be of two types: gas–liquid chromatography (GLC) and gas–solid chromatography.

17.3.2 PRINCIPLE

In GLC the separation is brought about by partitioning the sample between a mobile gas phase and a thin nonvolatile liquid layer coated on some inert solid particles, while gas–solid chromatography is based upon selective adsorption of constituents of the sample on a solid of large surface area used as the stationary phase.

When a mixture of volatile material transported by a carrier gas is led through a column containing an adsorbent solid phase or more commonly an absorbing liquid phase coated over a solid material, each volatile component is partitioned between the carrier gas and the solid or the liquid.

Depending upon the retention time in the column, the volatile components emerge from the column at different times and are finally detected by a suitable detector. If the carrier gas used the rate of flow and the temperature of the column is kept constant, the retention time (ie, the time taken by each component of mixture to traverse through the column) for each constituent of the mixture will always be the same.

It usually shows a linear relationship with the boiling point of the compound and is a characteristic for the constituents concerned under the given set of conditions for a given column. Thus, it is possible to identify the compound from its characteristic retention time on a particular column and under a given set of conditions. Quantitative estimation can be carried out from the extent of peak area recorded by the detector recorder system.

Apparatus used for gas chromatography is a simple tube of about 4 mm in diameter and about 120 cm to many meters in length. It is made of stainless steel or glass and is usually bent or coiled so that it could be accommodated in a small space. The tube or the column is packed with particles of some suitable adsorbent or in the case of GLC the fixed phase is a nonvolatile liquid coated over some solid support (particles of diatomaceous earth, crushed fire bricks, etc.).

As mobile phase, the gases used may be argon, helium, nitrogen, or hydrogen – hydrogen is usually not preferred because of fire hazards. There has been a recent trend to use capillary-gas chromatography in which instead of the column a very thin capillary, about 0.25 mm in diameter and usually 50 m in length made of glass, stainless steel, or some organic polymer, is employed. The inside of the capillary wall is coated with the stationary liquid phase. These capillary columns are superior to packed columns in terms of separation efficiency. They can separate up to several components from a single sample.

To maintain a constant rate of flow of the carrier gas, there is a flow meter and an adjustment device that regulates the flow of the carrier gas into the column. The sample is introduced through a self-sealing silicon rubber partition into a chamber that is heated to bring about evaporation of the sample. The temperature of the chamber must not be so high as to decompose the sample.

Solid samples have to be dissolved in some solvent, whereas gaseous samples require special sample introduction valves. The detectors, placed at the exit of separation chamber, detect and measure the small amounts of separated components present in the stream of the carrier gas leaving the column. Normally, three types of detectors are employed in gas chromatography: thermal conductivity detectors, flame ionization detectors, and electron capture detectors.

17.3.3 DETECTORS

17.3.3.1 Thermal conductivity detector

Thermal conductivity detectors are the most widely used detectors in gas chromatography. These detectors use heated metal filament (or thermisters that are made of some semiconductor of fused metal oxides) to sense small changes in thermal conductivity of the carrier gas. Thermal conductivity of the carrier gas only gives an essentially constant signal. The presence of vapors of the different components of the mixture in carrier gas brings about changes in the thermal conductivity proportional to their amount in the stream. This brings about changes in the resistance of the filament which is measured. The recorder that records these changes is equipped with an automatic device that traces the magnitude of these changes on a graph sheet along with the retention time.

17.3.3.2 Flame ionization detector

Flame ionization detectors are based on the measurement of electrical conductivity of gases. At normal temperatures and pressure, gases act as a bad conductor or insulators but when ionized they act as a good conductor of electrical current.

Gases and vapors as they emerge from the separation column are mixed with hydrogen and burned in air to produce a flame which ionizes the component molecules in the carrier gas. The burning jet is the negative electrode, while the anode is usually a small loop of wire extending into the tip of the flame across which a small voltage is applied.

The ions produced are collected at the electrodes and a current is generated which is proportional to the number of the component molecules ionized. Only the carrier gas burning with hydrogen produces an essentially constant signal but when the components of the mixture being analyzed emerge ionization occurs and a higher current is observed. The detector is equipped with an automatic recording device that records these fluctuations and transmits them to a graph sheet along with the retention time.

17.3.3.3 Electron capture detector

Electron capture detector is based on the phenomenon of electron capture of compounds having an affinity for free electrons. A 3-ray source is used to obtain slow electrons by ionization of the carrier gas (nitrogen is preferred) passing through the detector.

These electrons as they flow toward the anode under a fixed potential give rise to a steady current. When the component molecules of the mixture being analyzed come out of the separation column and pass through the detector these electrons are trapped.

The net result is replacement of the electrons by negatively charged ions of much greater mass and corresponding reduction in the flow of electric current proportional to the concentration of electron capturing component in the carrier gas. These changes are detected by using a suitable circuit and recorded with the help of an automatic device that traces a graph along with the retention time.

17.4 HIGH-PERFORMANCE LIQUID CHROMATOGRAPHY (HPLC)

Determination of quality parameters under stress conditions using HPLC (Fig. 17.2)

Principle: Separation of mixture of compound into individual components through equilibrium distribution between stationery and mobile phases.

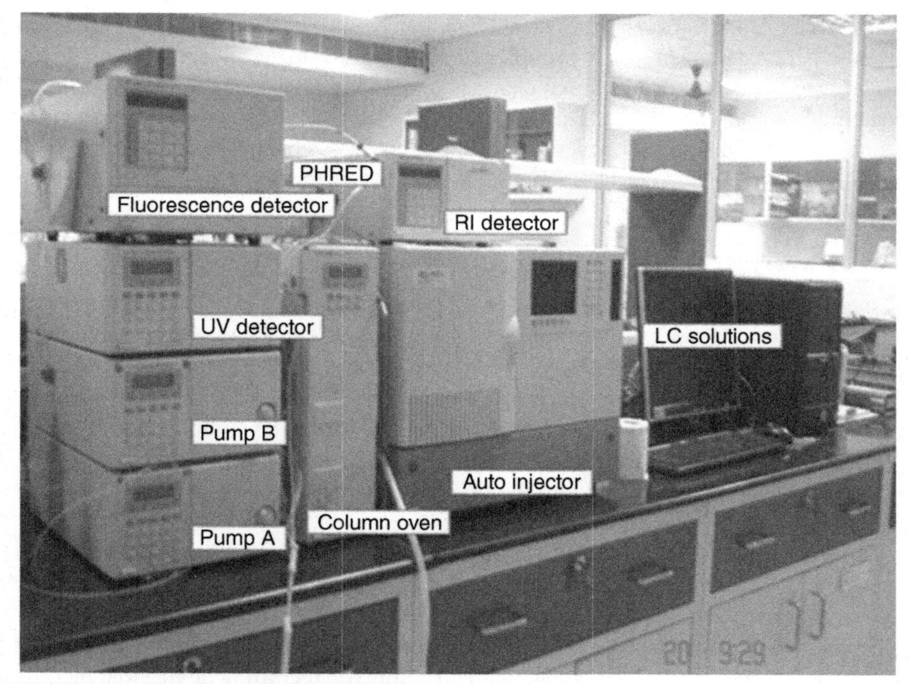

FIGURE 17.2 High-Performance Liquid Chromatography.

It is a separation technique that involves

- injection of a small volume of liquid sample
- into a tube (column) packed with tiny particles (3–5 μm in diameter called stationery phase)
- where individual components of the sample are moved down the packed tube (column) with a liquid (mobile phase) forced through the column by high pressure delivered by a pump.

These components are separated from one another by the column packing that involves various chemical and or physical interactions between their molecules and the packing particles. These components are detected at the exit of this tube (column) by a flow through device (detector) that measures their amount. An output from this detector is called a "liquid chromatogram."

17.4.1 ROLE OF FIVE MAJOR HPLC COMPONENTS

Pump: The role is to force a liquid (called the mobile phase) through the liquid chromatography at a specific flow rate expressed in milliliters per minute. The normal flow rates are 1–2 mL/min. Pumps can reach pressures in the range of 6000–9000 psi (400–600 bars).

Based on the requirement of one or two pumps, pumps are of two types:

1. *Isocratic pump*: It delivers constant mobile phase composition (solvent is premixed) and remains constant with time.
2. *Gradient pump*: It delivers variable mobile phase composition (binary gradient delivers two solvents). Mobile phase solvent composition increases with time.

Autosampler (injector): The injector serves to introduce the liquid sample into the flow stream of the mobile phase (sample volume ranges between 5 and 20 μL) automatically. The injector must also be able to withstand the high pressures of the liquid system.

Column: Column is considered the "heart of the chromatograph." Column stationery phase separates the sample components of interest by interaction between the sample components and the column packing material. The small particles inside the column cause high back pressure at normal flow rates. Analytical column internal diameter ranges from 1.0 to 4.6 mm and lengths from 15 to 250 mm. Columns are packed using high pressure to ensure that they are stable. There are two major separation modes that are used to separate most compounds.

1. In reverse phase chromatography, the column packing material is nonpolar (C18, C8, C3, phenyl, etc.) and the mobile phase is water (buffer) or water miscible organic solvents (eg, methanol, acetonitrile). The organic solvent increases the solvent strength and elutes compounds that are very strongly retained. The packing material in C18 column is octa dodecyl silane which has highly stable bonds.
2. In normal phase chromatography, the column packing is polar and the mobile phase is nonpolar (eg, hexane, isooctane, ethyl acetate).

Detectors: The detector can see (detect) the individual molecules that come out (elute) from the column and provides an output to a computer that results in the liquid chromatogram (ie, the graph of the detector response). It serves to analyze the individual molecules both qualitatively and quantitatively. There are many detection principles used to detect the compounds eluting from an HPLC column.

The most common are as follows:

1. *UV detection*: An ultraviolet light beam is directed through a flow cell and a sensor measures the light passing through the cell. The amount of light energy absorbed by the eluted compound from the column is measured by UV detector and aid in the identification of a compound or series of compounds.
2. *Fluorescence detection*: Compared to UV detectors, fluorescence detectors offer a higher sensitivity and selectivity that allows quantifying and identifying compounds at extremely low concentration levels. Fluorescence detectors sense only those substances that fluoresce.
3. *Refractive index detection (RI)*: The RI is a measure of molecule's ability to deflect light in a flowing mobile phase in a flow cell relative to a static mobile phase contained in a reference cell. The amount of deflection is proportional to concentration.

17.5 LIQUID CHROMATOGRAPHY–MASS SPECTROMETRY (LC–MS, OR ALTERNATIVELY HPLC–MS)

Liquid chromatography-mass spectrometry (LC–MS, or alternatively HPLC–MS) is an analytical chemistry technique that combines the physical separation capabilities of liquid chromatography (or HPLC) with the mass analysis capabilities of mass spectrometry (MS). LC–MS is a powerful technique that has very high sensitivity and selectivity and so is useful in many applications. Its application is oriented toward the separation, general detection, and potential identification of chemicals of particular masses in the presence of other chemicals (ie, in complex mixtures), for example, natural products from natural-products extracts, and pure substances from mixtures of chemical intermediates. Preparative LC–MS systems can be used for rapid mass-directed purification of specific substances from such mixtures that are important in basic research, and pharmaceutical, agrochemical, food, and other industries.

Present-day liquid chromatography generally utilizes very small particles packed and operating at relatively high pressure, and is referred to as HPLC; modern LC–MS methods use HPLC instrumentation, essentially exclusively, for sample introduction. In HPLC, the sample is forced by a liquid at high pressure (the mobile phase) through a column that is packed with a stationary phase generally composed of irregularly or spherically shaped particles chosen or derivatized to accomplish particular types of separations. HPLC methods are historically divided into two different subclasses on the basis of stationary phases and the corresponding required polarity of the mobile phase. Use of octadecylsilyl (C18) and related organic-modified particles as stationary phase with pure or pH-adjusted water–organic mixtures such as water–acetonitrile and water–methanol are used in techniques termed reversed phase liquid chromatography (RP-LC). Materials such as silica gel as stationary phase with neat or mixed organic mixtures are used in techniques termed normal phase liquid chromatography (NP-LC). RP-LC is most often used as the means to introduce samples into the MS, in LC–MS instrumentation.

17.5.1 FLOW SPLITTING

When standard bore (4.6 mm) columns are used the flow is often split ~10:1. This can be beneficial by allowing the use of other techniques in tandem such as MS and UV detection. However, splitting the flow to UV will decrease the sensitivity of spectrophotometric detectors. The mass spectrometry, on the other hand, will give improved sensitivity at flow rates of 200 μL/min or less.

17.5.2 MASS SPECTROMETRY (MS)

Mass spectrometry (MS) is an analytical technique that measures the mass-to-charge ratio of charged particles. It is used for determining masses of particles, for determining the elemental composition of a sample or molecule, and for elucidating the chemical structures of molecules, such as peptides and other chemical compounds. MS works by ionizing chemical compounds to generate charged molecules or molecule fragments and by measuring their mass-to-charge ratios. In a typical MS procedure, a sample is loaded onto the MS instrument and undergoes vaporization. The components of the sample are ionized by one of a variety of methods (eg, by impacting them with an electron beam), which results in the formation of charged particles (ions). The ions are separated according to their mass-to-charge ratio in an analyzer by electromagnetic fields. The ions are detected, usually by a quantitative method. The ion signal is processed into mass spectra.

Additionally, MS instruments consist of three modules: an ion source, which can convert gas phase sample molecules into ions (or, in the case of electrospray ionization, move ions that exist in solution into the gas phase); a mass analyzer, which sorts the ions by their masses by applying electromagnetic fields; and a detector, which measures the value of an indicator quantity and thus provides data for calculating the abundances of each ion present.

The technique has both qualitative and quantitative uses. These include identifying unknown compounds, determining the isotopic composition of elements in a molecule, and determining the structure of a compound by observing its fragmentation. Other uses include quantifying the amount of a compound in a sample or studying the fundamentals of gas phase ion chemistry (the chemistry of ions and neutrals in a vacuum). MS is now in very common use in analytical laboratories that study physical, chemical, or biological properties of a great variety of compounds.

17.5.3 MASS ANALYZER

There are many different mass analyzers that can be used in LC/MS: single quadrupole, triple quadrupole, ion trap, time of flight (TOF), and quadrupole-time of flight.

17.5.4 INTERFACE

Understandably, the interface between a liquid phase technique that continuously flows liquid, and a gas phase technique carried out in a vacuum, was difficult for a long time. The advent of electrospray ionization changed this. The interface is most often an electrospray ion source or variant such as a nanospray source; however, atmospheric pressure chemical ionization interface is also used. Various deposition and drying techniques have also been used such as using moving belts; however, the most common of these is off-line MALDI deposition. A new approach still under development called Direct-EI LC-MS interface couples a nano HPLC system and an electron ionization equipped mass spectrometer (Stobiecki et al., 2006).

17.5.5 APPLICATIONS

17.5.5.1 Pharmacokinetics

LC–MS is very commonly used in pharmacokinetic studies of pharmaceuticals and is thus the most frequently used technique in the field of bioanalysis. These studies give information about how quickly a drug will be cleared from the hepatic blood flow, and organs of the body. MS is used for this due to high sensitivity and exceptional specificity compared to UV (as long as the analyte can be suitably ionized), and short analysis time.

The major advantage MS has is the use of tandem MS-MS. The detector may be programmed to select certain ions to fragment. The process is essentially a selection technique, but is in fact more complex. The measured quantity is the sum of molecule fragments chosen by the operator. As long as there are no interferences or ion suppression, the LC separation can be quite quick.

17.5.5.2 Proteomics/metabolomics

LC–MS is also used in proteomics where again components of a complex mixture must be detected and identified in some manner. The bottom-up proteomics LC–MS approach to proteomics generally involves protease digestion and denaturation (usually trypsin as a protease, urea to denature tertiary structure, and iodoacetamide to cap cysteine residues) followed by LC–MS with peptide mass fingerprinting or LC–MS/MS (tandem MS) to derive sequence of individual peptides. LC–MS/MS is most commonly used for proteomic analysis of complex samples where peptide masses may overlap even with a high-resolution mass spectrometer. Samples of complex biological fluids such as human serum may be run in a modern LC–MS/MS system and result in over 1000 proteins being identified, provided that the sample was first separated on an SDS-PAGE gel or HPLC-SCX.

Profiling of secondary metabolites in plants or food like phenolics can be achieved with liquid chromatography–mass spectrometry.

17.6 INDUCTIVELY COUPLED PLASMA SPECTROMETRY (ICP) (SOIL & PLANT ANALYSIS LABORATORY UNIVERSITY OF WISCONSIN–MADISON HTTP://UWLAB.SOILS.WISC.EDU)

17.6.1 INTRODUCTION

Inductively coupled plasma spectrometry is a type of mass spectrometry which is capable of detecting metals and several nonmetals at concentrations as low as one part in 1015 (part per quadrillion, ppq) on noninterfered low-background isotopes. This is achieved by ionizing the sample with inductively coupled plasma and then using a mass spectrometer to separate and quantify those ions.

Compared to atomic absorption techniques, ICP has greater speed, precision, and sensitivity. However, compared with other types of mass spectrometry, such as TIMS and Glow Discharge, ICP introduces a lot of interfering species: argon from the plasma, component gasses of air that leak through the cone orifices, and contamination from glassware and the cones.

Analysis of major, minor, and trace elements in plant tissue samples can be done by inductively coupled plasma optical emission spectrometry (ICP-OES) and inductively coupled plasma mass spectrometry (ICP-MS). It allows determination of elements with atomic mass ranges 7–250 (lithium to uranium), and sometimes higher. ICP is also used widely in the geochemistry for radiometric dating, in which it is used to analyze relative abundance of different isotopes, in particular uranium and lead. In the pharmaceutical industry, ICP is used for detecting inorganic impurities in pharmaceuticals and their ingredients. One of the largest volume uses for ICP is in the medical and forensic field, specifically, toxicology. In recent years, industrial and biological monitoring has presented another major need for metal analysis via ICP-MS.

17.6.2 SUMMARY OF METHOD

Half a gram of dried sample (or equivalent) and 5 mL of concentrated nitric acid are added to a 50-mL Folin digestion tube. The mixture is heated to 120–130°C for 14–16 h and is then treated with hydrogen peroxide. After digestion, the sample is diluted to 50 mL. This solution is analyzed by ICP-OES for major and minor components, and further 1:1 diluted and analyzed by ICP-MS for minor and trace components.

After solid samples are converted into solution samples, the procedures of "elemental analysis of solution samples with ICP-OES" and "elemental analysis of solution samples with ICP-MS" are followed.

17.6.3 SAFETY

All chemicals should be considered potential health hazard. All relevant laboratory safety procedures are followed.

The use of perchloric acid for a sample digestion must be conducted in a hood designed specifically for perchloric acid. The user must be aware of the dangers involved using perchloric acid, such as the explosive nature of anhydrated perchloric acid and its extreme corrosive nature.

17.6.4 INTERFERENCE

This method covers the analysis of over 30 elements in different kinds of samples by ICP-OES and ICP-MS. A general discussion of interference is lengthy but not necessarily relevant to a specific element, which is especially true if the sample matrix is not specifically defined. An enormous amount of literature is available to the analysis of metals and nonmetals by ICP-OES and by ICP-MS. Reading the published articles is recommended.

In this method, the solution contains less than 1000 ppm of dissolved solids for ICP-OES and ICP-MS analysis. The major components are K, Mg, Ca, P, S, and Na. These components either do not pose significant interferences with other elements/isotopes or the potential interferences are well understood and controlled. Significant interferences are not expected, although some specific elements and or isotopes may be interfered.

17.6.5 MEASUREMENT BY ICP-OES

17.6.5.1 Sample preparation

Set 8-mL autosampler tubes in ICPOES sample racks. Transfer sample solutions from 50-mL tubes to 8-mL tubes. For samples with extremely high analytes, the samples may be further diluted. Add 3 mL of sample solution and 3 mL of 2% nitric acid to the 8-mL autosampler tube (2nd dilution. Nominal dilution factor = 200. $Y = 4$ ppm).

17.6.6 MEASUREMENT

A detailed procedure is given in "elemental analysis of solution samples with ICP-OES." Digestion blanks are also measured with other samples.

17.6.7 MEASUREMENT BY ICP-MS

17.6.7.1 Sample preparation

Set 14-mL Falcon tubes in the ICPMS autosampler racks. Transfer the sample solutions to the Falcon tubes. Adjust the volume to 5 mL. Add 5 mL of 2% nitric acid. Mix well. The nominal dilution factor is 200 and the IRS is 4 ppb of Rh. Since an internal reference standard is used, the volume inaccuracy during dilution is irrelevant.

If the concentrations of target elements are expected to be relatively high, the samples are further diluted, by either 2 + 8 dilution or 1 + 9 dilution. Otherwise, a sample solution may be directly analyzed without any further dilution (ie, 10 + 0 dilution). During the data processing in later stage, the nominal dilution factor is always 200, whether the dilution is 1 + 9, 2 + 8, 5 + 5, or 10 + 0.

17.6.8 MEASUREMENT

A detailed procedure is given in "elemental analysis of solution samples with ICP-MS." Edit the menu depending on specific samples or analytical requests.

Appendices

COMMON BUFFERS

In this chapter, we describe the methods of preparation of some of the buffers, most commonly used in the assay of enzymes required for plant physiological and histochemical studies.

The buffers have been arranged in an alphabetical order starting with their names. These methods are not necessarily identical to those of the original authors. It is advised that the users should redetermine the titration curves of majority of the buffers. These buffers have also been listed in Method in Enzymology, Vol. 1 (1955), pp. 138–146.

APPENDIX I: CITRATE BUFFER

Stock solution

A : 0.1 M solution of citric acid (21.01 g in 1000 mL water)

B : 0.1 M solution of sodium citrate (29.41 g of $C_6H_5O_7Na_32H_2O$ in 1000 mL water).

x mL of A + *y* mL of B, diluted to a total of 100 mL.

x	*y*	pH
46.5	3.5	3.0
43.7	6.3	3.2
40.0	10.0	3.4
37.0	13.0	3.6
35.0	15.0	3.8
33.0	17.0	4.0
31.5	18.5	4.2
28.0	22.0	4.4
25.5	24.5	4.6
23.0	27.0	4.8
20.0	29.5	5.0
18.0	32.0	5.2
16.0	34.0	5.4
13.7	36.3	5.6
11.8	38.2	5.5
9.5	41.5	6.0
7.2	42.8	6.2

Phenotyping Crop Plants for Physiological and Biochemical Traits. http://dx.doi.org/10.1016/B978-0-12-804073-7.00019-3

APPENDIX II: SODIUM PHOSPHATE BUFFER

Stock solutions

A: 0.2 M solution of monobasic sodium phosphate (27.8 g in 1000 mL water).

B: 0.2 M solution of dibasic sodium phosphate (53.65 g of $Na_2HPO_4.7H_2O$ or 71.7 g of $Na_2.HPO_4$, $12H_2O$ in 1000 mL).

x mL of A + y mL of B, diluted to a total of 200 mL.

x	*y*	pH	*x*	*y*	pH
93.5	6.5	5.7	45.0	55.0	6.9
92.0	8.0	5.8	39.0	61.0	7.0
90.0	10.0	5.9	33.0	67.0	7.1
87.7	12.3	6.0	28.0	72.0	7.2
85.0	15.0	6.1	23.0	77.0	7.3
81.5	18.5	6.2	19.0	81.0	7.4
77.5	22.5	6.3	16.0	84.0	7.5
73.5	26.5	6.4	13.0	87.0	7.6
68.5	31.5	6.5	10.5	90.5	7.7
62.5	37.5	6.6	8.5	91.5	7.8
56.5	43.5	6.7	7.0	93.0	7.9
51.0	49.0	6.8	5.3	94.7	8.0

APPENDIX III: POTASSIUM PHOSPHATE BUFFER

pH	% K_2HPO_4 (Dibasic)	% KH_2PO_4 (Monobasic)
5.8	8.5	91.5
6.0	13.2	86.8
6.2	19.2	80.8
6.4	27.8	72.2
6.6	38.1	61.9
6.8	49.7	50.3
7.0	61.5	38.5
7.2	71.7	28.3
7.4	80.2	19.8
7.6	86.6	13.4
7.8	90.8	9.2
8.0	94.0	6.0

APPENDIX IV: SODIUM ACETATE BUFFER

Stock solutions

A: 0.2 M solution of acetic acid (11.55 mL in 1000 mL water).

B: 0.2 M solution of sodium acetate (16.4 g of $C_2H_3O_2Na$ or 27.2 g of $C_2H_3O_2Na$-$3H_2O$ in 1000 mL water).

x mL of A+ *y* mL of B, diluted to a total of 100 mL.

x	*y*	pH
46.3	3.7	3.6
44.0	6.0	3.8
41.0	9.0	4.0
36.8	13.2	4.2
30.5	19.5	4.4
25.5	24.5	4.6
20.0	30.0	4.8
14.8	35.2	5.0
10.5	39.5	5.2
8.8	41.2	5.4
4.8	45.2	5.6

APPENDIX V: TRIS–HCl BUFFER (TRIS–HYDROXYMETHYL AMINOMETHANE HYDROCHLORIC ACID BUFFER)

Stock solutions

A: 0.2 M solution of (Tris-hydroxymethyl aminomethane) (24.2 g in 1000 mL water).

B: 0.2 M HCl

50 mL of A + *x* mL of B, diluted to a total of 200 mL.

X	pH
5.0	9.0
8.1	8.8
12.2	8.6
16.5	8.4
21.9	8.2
26.8	8.0
32.5	7.8
38.4	7.6
41.4	7.4
44.2	7.2

APPENDIX VI: 1M HEPES–NaOH pH 7.5 BUFFER

700 mL D.D.H_2O, 238.3 g HEPES, and 5.5 g NaOH pellets added to adjust pH to 7.5.

APPENDIX VII: PREPARATION OF STOCKS OF MACRO AND MICRONUTRIENTS FOR HYDROPONICS EXPERIMENT

Macronutrients Stocks	Molecular Weight (g)	Concentration of Stock Solution (1 M)	Concentration of Stock Solution (g/L)	Volume of Stock Solution/L of Final Solution (mL)	Element	μM	ppm
KNO_3	101.10	1.00	101.10	6.0	N	16,000	224
					K	6000	235
$Ca(NO_3)_2.4H_2O$	236.16	1.00	236.16	4.0	Ca	4000	160
$NH_4H_2PO_4$	115.08	1.00	115.08	2.0	P	2000	62
$MgSO_4$	246.49	1.00	246.49	1.0	S	1000	32
$NaNO_3$	84.99	1.00	84.99	1.0	Mg	1000	24

Micronutrients Stock	Molecular Weight (g)	Concentration of Stock Solution (mM)	Concentration of Stock Solution (g/L)	Volume of Stock Solution/L of Final Solution (mL)	Element	μM	ppm
KCl	74.55	50	3.728	1.0	Cl	50	1.77
			1.546				
H_3BO_3	61.84	25	0.338		B	25	0.27
$MnSO_4.H_2O$	169.01	2.0	0.575		Mn	2.0	0.11
$ZnSO_4.7H_2O$	287.55	2.0	0.125		Zn	2.0	0.131
$CuSO_4.5H_2O$	249.71	0.5	0.081		Cu	0.5	0.032
$(NH_4)_6\ MO_7O_{24}.4H_2O$	1235.9	0.5	1 L		Mo	0.5	0.050
Fe-EDTA	-	-	-	1.0	Fe	-	5.00

Preparation of Fe-EDTA: Use amber color bottle for Fe-EDTA preparation. Dissolve 26.1 g of EDTA in 286 mL of 1N KOH and mix with 24.9 g Ferrous sulfate (FeSO4.7H2O) and diluted to 1 L. Keep overnight aeration for getting available state of Fe. One milliliter of this solution provided 5 ppm in 1 L.

APPENDIX VIII: PREPARATION OF 'HOAGLAND SOLUTION' FOR HYDROPONICS EXPERIMENT

Stock Solutions	Stocks Full Strength, mL/L											
	Com-plete	-N	-P	-K	-Ca	-Mg	-S	-Fe	-Zn	-Mn	-B	-Cu
KNO_3	6	-	6	-	6	6	6	6	6	6	6	6
$Ca(NO_3)_2 \cdot 4H_2O$	4	-	4	4	-	4	4	4	4	4	4	4
$NH_4H_2PO_4$	2	-	-	2	2	2	2	2	2	2	2	2
$MgSO_4$	1	1	1	1	1	-	-	1	1	1	1	1
$NaNO_3$ (M.W:85.0)	-	-	-	6	8	-	-	-	-	-	-	-
Fe-EDTA	1	1	1	1	1	1	1	-	1	1	1	1
Micronutrients	1	1	1	1	1	1	1	1	1 (-Zn)	1 (-Mn)	1 (-B)	1 (-Cu)
$MgCl_2\ 6H_2O$ (M.W:95.23)	-	-	-	-	-	-	1@	-	-	-	-	-
Na_2SO_4 (M.W.142.05)	-	-	-	-	-	1	-	-	-	-	-	-
$CaCl_2$ (M.W.110.99)	-	4*	-	-	-	$ 1	-	-	-	-	-	-
KCl (M.W.74.55)	-	6*	-	-	-	-	-	-	-	-	-	-
$NaH_2PO_4 \cdot 2H_2O$ (M.W:156.01)	-	2**	-	-	-	-	-	-	-	-	-	-
NH_4Cl (M.W:53.46)	-	-	2***	-	-	-	-	-	-	-	-	-

**355 ppm of Cl; **46 ppm of Na; ***71 ppm of Cl; @ 35 ppm of Cl; $ 23 ppm of Na*

APPENDIX IX: SOLUBILITY CHART OF PLANT GROWTH REGULATORS

Sr. No.	Growth Regulator	Solubility
1	ABA	1 N NaOH
2	6-BAP	1 N NaOH
3	CCC	Water
4	2,4 D	Ethyl alcohol/1 N NaOH
5	GA_3	Ethyl alcohol
6	IAA	Ethyl alcohol/1 N NaOH
7	Kinetin	1 N NaOH
8	Maleic hydrazide	1 N NaOH
9	NAA	1 N NaOH

References

A.O.A.C., 1960. Official and tentative method of analysis A.O.A.C. Washington, DC.

Abdul-Baki, A.A., Anderson, J.D., 1973. Vigour determination in soybean seed by multiple criteria. Crop Sci. 13, 630–633.

Albrecht, J.A., 1993. Acorbic acid and retention in lettuce. J. Food Qual. 16, 311–316.

Araus, J.L., Slafer, G.A., Reynolds, M.P., 2002. Plant breeding and drought in C3 cereals: what should we breed for? Ann. Botany 89, 925–940.

Arnon, D.I., 1949. Copper enzymes in isolated chloroplasts: polyphenol oxidases in Beta vulgaris. Plant Physiol. 24, 1–14.

Asada, K., 1994. Production and action of active oxygen species in photosynthetic tissue. In: Foyer, C.H., Mullineaux, P.M. (Eds.), Causes of Photooxidative Stress Amelioration of Defense Systems in Plants. CRC Press, Boca Raton, FL, pp. 77–104.

Babitha, M., Sudhakar, P., Latha, P., Reddy, P.V., 2006. Screening of groundnut genotypes for high temperature tolerance. Plant Physiol. 11 (1), 63–74.

Balasubramanian, V., Morales, A.C., Cruz, R.T., Thiyagarajan, T.M., Nagarajan, R., Babu, M., Abdulrachman, S., Hai, L.H., 2000. Adaptation of the chlorophyll meter (SPAD) technology for real time N management in rice: a review. Int. Rice Res. Notes 25 (1), 4–8.

Balasubramanian, T., Sadasivam, S., 1987. Changes in starch, oil, protein and amino acids in developing seeds of okra (*Abelmoschus esculentus* L. Moench). Plant Foods for Human Nutrition 37, 41–46.

Barbour, M.M., Farquhar, G.D., 2000. Relative humidity and ABA-induced variation in carbon and oxygen isotope ratios of cotton leaves. Plant Cell Environ. 23, 473–485.

Barrs, H.D., Weatherlay, P.E., 1962. A re-examination of the relative turgidity technique for estimating water deficit in leaves. Austral. J. Biol. Sci. 15, 413–428.

Baskin, C.C., 1969. GADA and seedling at tests for seed quality. Proc. Seedsmaen Shot Commun. Mississipi State Univ.: 59–69.

Bates, L., Waldren, R.P., Teare, I.D., 1973. Rapid determination of free proline water stress studies. Plant Soil 39, 205–207.

Beauchamp, C., Fridovich, I., 1971. Superoxide dismutase: improved assays and an assay applicable to acrylamide gels. Anal. Biochem. 44 (1), 276–287.

Begum, H., Lavanya, M.L., Ratna Babu, G.H.V., 1987. Effect of pre-sowing treatments on seed and seedling vigour in papaya. Seed Res. 15, 9–15.

Belanger, J., Balakrishna, M. Latha, P., Katumalla, S., Johns, T., 2010. Contribution of selected wild and cultivated leafy vegetables from South India to lutein and β carotene intake. Asia Pacific J. Clin. Nutr. 19 (3), 417–424.

Bindu Madhava, H., Sheshshayee, M.S., Devendra, R., Prasad, T.G., Udayakumar, M., 1999. Oxygen (^{18}O) isotope enrichment in the leaves as a potential surrogate for transpiration and stomatal conductance. Curr. Sci. 76 (11), 1427–1428.

Bino, R.J., Aartse, J.W., Van Der Burg, W.J., 1993. Non-destructive X-ray analysis of Arabidopsis embryo mutants. Seed Sci. Res. 3, 167–170.

Blackman, G.E., 1968. The application of the concepts of growth analysis to the assessment of productivity. In: Eckardt, F.E. (Ed.), Functioning of Terrestrial Ecosystems at the Primary Production Level. UNESCO, Paris.

Phenotyping Crop Plants for Physiological and Biochemical Traits. http://dx.doi.org/10.1016/B978-0-12-804073-7.00024-7

Blackman, V.H., 1919. The compound interest law of plant growth. Ann. Botany 33, 353–360.
Blum, A., 2005. Drought resistance, water-use efficiency and yield potential: are they compatible, dissonant or mutually exclusive? Austral. J. Agric. Res. 56, 1159–1168.
Borlaug, N.E., 2007. Sixty-two years of fighting hunger: personal recollections. Euphytica 157, 287–297.
Borrell, A.K., Hammer, G.L., 2000. Crop Sci. 40, 1295–1307.
Bowler, C., Van Montague, M., Inze, D., 1992. Superoxide dismutase and stress tolerance. Annu. Rev. Plant Physiol. Molecular Biol. 43, 83–116.
Bradford, M.M., 1976. A rapid and sensitive method for quantitation or microgram quantities of protein utilizing the principle of protein-dye binding. Anal. Biochem. 72, 248–254.
Briggs, G.E., Kidd, F., West, C., 1920. A quantitative analysis of plant growth. Ann. Appl. Biol. 7 (2–3), 202–223.
Bulingame, A.L., Boyd, R.K., Gaskell, S.J., 1998. Mass spectrometry. Anal. Chem. 70 (16), 647–716.
Bullen, W.A., 1956. The isolation and characterization of glutamic dehydrogenase from corn leaves. Arch. Biochem. Biophys. 62, 173–183.
Chaitanya, K.S.K., Naithani, S.C., 1994. Role of superoxide, lipid peroxidation and superoxide dismutase is membrane perturbation during loss of variability in seeds of Shorea robusta. Faer – in F. New Phytol. 126, 623–627.
Chapman, S.C., Barreto, H.J., 1997. Using a chlorophyll meter to estimate specific leaf nitrogen of tropical maize during vegetative growth. Agronomy J. 89, 557–562.
Chen, G.-X., Asada, K., 1990. Hydroxyurea and p-aminophenol are the suicide inhibitors of ascorbate peroxidase. J. Biol. Chem. 265, 2775–2781.
Chinard, F.P., 1952. Photometric estimation of proline and ornithine. J. Biol. Chem. 199, 91–95.
Chung, H.J., Ferl, R.J., 1999. Arabidopsis alcohol dehydrogenase expression in both shoots and of roots is conditioned by root growth environment. Plant Physiol. 121, 429–436.
Condon, A.G., Richards, R.A., Rebetzke, G.J., Farquhar, G.D., 2004. Breeding for high water-use efficiency. J. Exp. Botany 55, 2447–2460.
Cooper, A.J., 1979. The ABC of NFT. Grower Books, London.
Craig, L., Gordon, L.I., 1965. Deuterium and oxygen-18 variations in the ocean and the marine atmosphere. In: Tongiorgi, E. (Ed.), In: Proceedings of a Conference on Stable Isotopes in Oceanographic Studies and Paleotemperatures. Spoleto, Italy, pp. 9–130.
Cushman, J.C., Bohnert, H.J., 2000. Genomic approaches to plant stress tolerance. Curr. Opinion Plant Biol. 3, 117–124.
Dadlani, M., Agrawal, P.K., 1983. Factors influencing leaching of sugars and electrolytes from carrot and okra seeds. Sci. Hortic. 19, 39–44.
Dadlani, M., Agrawal, P.K., 1987. Determination of lipid peroxidation. In: Agrawal, P.K., Dadlani, M. (Eds.), Techniques in Seed Science Technology. South Asian Publishers, New Delhi, p. 129.
Deak, M., Horvath, G.V., Davletova, S., Torok, K., Sass, L., Vass, I., Barna, B., Kiraly, Z., Dudits, D., 1999. Plants ectopically expressing the iron-binding protein, ferritin, are tolerant to oxidative damage and pathogens. Nat. Biotechnol. 17, 192–196.
Deepa, M., Rama Rao, G., Latha, P., Naidu, M.V.S., 2012. Physiological evaluation of greengram genotypes under moisture stress conditions. Andhra Agric. J. 59 (1), 276–279.
Dell'Aquila, A., 2006. Computerised seed imaging: a new tool to evaluate germination quality. Commun. Biometry Crop Sci. 1 (1), 20–31.

Dwyer, L., Anderson, M.A.M., Ma, B.L., Stewart, D.W., Tollenaar, M., Gregorich, E., 1995. Quantifying the non-linearity in chlorophyll meter response to corn leaf nitrogen concentration. Canad. J. Plant Sci. 75, 179–182.

Ebercon, A.A., Blum, A., Jordan, W.R., 1977. A rapid calorimetric method for epicuticular wax content on sorghum leaves. Crop Sci. 17, 178–180.

Ehleringer, J.R., Osmond, C.B., 1988. Stable isotopes. In: Rundel, P.W., Ehleringer, J.R., Nagy, K.A. (Eds.), Stable Isotopes in Ecological Research. Springer-Verlag, New York, USA, pp. 281–299.

Falconer, D.S., 1981. In: Longman, L. (Ed.), Introduction to Quantitative Genetics, second ed. London, UK, p. 340.

Farquhar, G.D., Ehleringer, J.R., Hubick, K.T., 1989a. Carbon isotope discrimination and photosynthesis. Annu. Rev. Plant Physiol. Plant Mol. Biol. 40, 503–537.

Farquhar, G.D., Hubick, K.T., Condon, A.G., Richards, R.A., 1989b. Carbon isotope fractionation and plant water-use efficiency. In: Rundel, P.W., Ehleringer, J.R., Nagy, K.A. (Eds.), Stable Isotopes in Ecological Research. Springer-Verlag, New York, USA, pp. 21–40.

Farquhar, G.D., O'Leary, M.H., Berry, J.A., 1982. On the relationship between carbon isotope discrimination and the intercellular CO2 concentration in leaves. Austral. J. Plant Physiol. 9, 121–131.

Fender, S.E., O'Connell, M.A., 1990. Expression of heat shock response in a tomato interspecific hybrid is not intermediate between the two parental responses. Plant Physiol. 93, 1140–1146.

Fischer, K.S., Wood, 1981. Breeding and selection for drought resistance in tropical maize. In: Fischer, K.S., Jonson, E.C., Edmeades, G.O. (Eds.), In: Proceedings of Symposium on Principles and Methods in Crop Improvement for Drought Resistance with Emphasis on Rice. IRRI, Phillipines, p. 1981.

Fischer, R.A., Turner, N.C., 1978. Plant productivity in the arid and semiarid zones. Annu. Rev. Plant Physiol. 29, 277–317.

Flanagan, L.B., 1993. Environmental and biological influence on the stable oxygen and hydrogen isotopic composition of leaf water. In: Ehleringer, J.R., Hall, A.E., Farquhar, G.D. (Eds.), Stable Isotopes Plant Carbon–Water Relations. Academic Press, San Diego, CA, pp. 71–90.

Fokar, M., Nguyen, H.T., Blum, A., 1998. Heat tolerance in spring wheat: I. Estimating cellular thermotolerance and its heritability. Euphytica 104, 1–8.

Foyer, C.H., Lelandais, M., Kunert, K.J., 1994. Photooxidative stress in plants. Physiol. Plantarum 92, 696–717.

Fullner, K., Temperton, V.M., Rascher, U., Jahnke, S., Rist, R., Schurr, U., Kuhn, A.J., 2012. Vertical gradient in soil temperature stimulates development and increases biomass accumulation in barley. Plant Cell Environment. In press. doi: 10.1111/j.1365-3040.2011.02460.x.

Garay, A.F., Wilhelm, W.W., 1983. Root system characteristics of two soybean isolines undergoing water stress conditions. Agron. J. 75, 973–977.

Garnier, E., Freijsen, A.H.J., 1994. On ecological inference from laboratory experiments conducted under optimum conditions. In: Roy, J., Garnier, E. (Eds.), A Whole-Plant Perspective on Carbon-Nitrogen Interactions. SPB Academic Publishing, The Hague, The Netherlands, pp. 267–292.

George, W.S., Roshie, B., Keeling, P.L., 1994. Heat stress during grain filling in maize: effects on carbohydrate storage and metabolism. Aust. J. Plant Physiol. 21, 829–841.

Gorbe, E., Calatayud, A., 2010. Optimization of nutrition insoilless systems: a review. Adv. Bot. Res. 53, 193–245.

Graham, R.D., Welch, R.M., 1996. Breeding for staple-foodcrops with high micronutrient density. Int. Food Policy Res. Inst. Washington, DC.
Granier, C., Aguirrezabal, L., Chenu, K., Cookson, S.J., Dauzat, M., Hamard, P., Thioux, J.J., Rolland, G., Bouchier-Combaud, S., Lebaudy, A., Muller, B., Simonneau, T., Tardieu, F., 2006. PHENOPSIS, an automated platform for reproducible phenotyping of plant responses to soil water deficit in Arabidopsis thaliana permitted the identification of an accession with low sensitivity to soil water deficit. New Phytol. 169, 623–635.
Greive, C.M., Grattan, S.R., 1983. Rapid assay for determination of water-soluble quaternary amino compounds. Plant Soil 70, 303–307.
Hammer, G.L., Wright, G.C., 1994. Aust. J. Agric. Res. 45, 575–589.
Hartree, E.F., 1955. Haematin compounds. In: Analysis, K., Peach, M.V., Traccy (Eds.), Modern Methods of Plant. Springer-Verlag, Berlin, pp. 197–245.
Havaux, M., 1993. Rapid photosynthetic adaptation to heat stress triggered in potato leaves by moderately elevated temperatures. Plant Cell Environ. 16, 461–467.
Hedge, J.E., Hofreiter, B.T., 1962. In: Whistler, R.L., Be Miller, J.N. (Eds.), Carbohydrate Chemistry, vol. 17, Academic Press, New York.
Hewitt, E.J., 1966. Sand and water culture methods used in the study of plant nutrition. Technical Communication no. 22. Commonwealth Bureau of Horticulture and Plantation Crops, Great Britain.
Heydecker, W., 1974. Germination of an idea. The priming of seeds. University of Nottingham School of Agriculture, Report Part III, 1968–1969 pp. 96–100.
Hiscox, J.D., Israelstam, G.F., 1979. A method for extraction of chlorophyll from leaf tissue without maceration. Can. J. Bot. 57, 1332–1334.
Hoagland, D.R., Snijder, W.C., 1933. Nutrition of strawberry plants under controlled conditions. Proc. Am. Soc. Hortic. Sci. 30, 288–296.
Hodge, J.E., Hofreiter, B.T., 1962. In: Whistler, R.L., Miller, J.N. (Eds.), Methods in Carbohydrate Chemistry. Academic Press, New York.
Huber, S.C., Bickoff, D.M., 1984. Evidence for control of carbon partitioning by fructose-2,6-bisphosphate in spinach leaves. Plant Physiol. 74, 445–447.
Hubick, K.T., Farquhar, G.D., 1989. Carbon isotope discrimination and ratio of carbon gained to water lost in barley cultivars. Plant Cell Environ. 12, 795–804.
Hubick, K.T., Farquhar, G.D., Shorter, R., 1986. Correlation between water-use efficiency and carbon isotope discrimination in diverse peanut germplasm. Austral. J. Plant Physiol. 13, 803–816.
Hussey, A., Long, S.P., 1982. Seasonal changes in weight of above and below-ground vegetation and dead plant material in a salt marsh at colne point. Essex. J. Ecol. 70, 752–772.
Impa, S.M., Nadaradjan, S., Sheshshayee, M.S., Bindumadahava, H, Prasad, T.G., Udayakumar, M., 2003. RAPD markers and stable isotope ratios to delineate the stomatal and mesophyll control of WUE in rice. In: Second International Congress of Plant Physiology, 8–12 January 2003, New Delhi, India.
ISTA International Seed Testing Association (2005). International rules for seed testing. International Seed Testing Association, Basser- dorf, Switzerland.
I.S.T.A. International Seed Testing Association, 1993. International rules for seed testing. Seed Sci. Technol. 21, 363.
Inze, D., Van Montague, M., 1995. Oxidative stress in plants. Curr. Opinion Biotechnol. 6, 153–158.

Johns, T., 2007. Agrobiodiversity, diet and human health. In: Jarvis, D.I., Padoch, C., Cooper, H.D. (Eds.), Managing Biodiversity in Agricultural Ecosystems. Columbia University Press, New York, p. 492.

Katiyar, S., Thakur, V., Gupta, R.C., Sarin, S.K., Das, B.C., 2000. P53 tumour suppressor gene mutations in heptacellur carcinoma patients in India, pp. 1565–1573.

Kittock, P.A., Law, A.G., 1968. Relationship of seedling vigour to respiration and tetrazolium chloride reduction of germinating wheat seeds. Agron. J. 60, 286–288.

Klepper, L., Flesher, D., Hageman, R.H., 1971. Generation of reduced nicotinamide adenine dinucleotide for nitrate reduction in green leaves. Plant Physiol. 48, 580–590.

Knegt, E., Bruinsma, J., 1973. A rapid, sensitive and accurate determination of indolyl-3-acetic acid. Phytochem 12, 753–756.

Kobza, J., Seemann, J.R., 1988. Mechanisms for light-dependent regulation of ribulose-1,5-bisphosphate carboxylase activity and photosynthesis in intact leaves. Proc. Natl. Acad. Sci. USA 85, 3815–3819, 22.

Kobza, J., Seeman, J.R., 1988. Mechanisms for light dependent regulation of ribulose-1,5-bisphosphate carboxylase activity and photosynthesis in intact leaves. Proc. Nat. Acad. Sci. USA 85, 3815, 3819.

Kong, W.W., Zhang, C., Liu, F., Nie p, C., He, Y., 2013. Rice seed cultivar identification using near infrared hyperspectral imaging and multivariate data analysis. Sensors 13, 8916–8927.

Kranner, L., Kastberger, G., Hartbauer, M., Pritchard, H.W., 2010. Noninvasive diagnosis of seed viability using infrared thermography. Proc. Natl. Acad. Sci. USA 107 (8), 3912–3917.

Krishnan, M., Nguyen, H.T., Burke, J.J., 1989. Heat shock protein synthesis and thermotolerance in wheat. Plant Physiol. 90, 140–145.

Krishnaveni, S., Balasubramanian, T., Sadasivam, S., 1984. Food Chem. 15, 229.

Latha, P., Sudhakar, P., Bala Krishna, M., Rajiya Begum, C., Raja Reddy, K., 2011. Estimation of groundnut kernel aflatoxins by high performance liquid chromatography using immunoaffinity column clean up and post column photochemical derivatization. Legume Res. 34 (1), 31–35.

Latha, P., Reddy, P.V., 2005. Determination of water use efficiency in groundnut by gravimetric method and its association with physiological parameters. Indian J. Plant Physiol. 10 (4), 322–326.

Latha, P., Reddy, P.V., 2007. Water use efficiency and its relation to specific leaf area carbon isotope discrimination and total soluble proteins under mid-season moisture stress conditions in groundnut, *Arcachis hypogaea* L. genotypes. J. Oilseeds Res. 24 (1), 77–80.

Latha, P., Sudhakar, P., Balakrishna, M., Raja Reddy, K., 2011. Survey on level of aflatoxin contamination in eastern and western mandals of chittoor district of Andhra Pradesh. J. Res., ANGRAU 39 (1 & 2), 34–36.

Latha, P., Sudhakar, P., Sreenivasulu, Y., Naidu, P.H., Reddy, P.V., 2007. Relationship between total phenols and aflatoxin production of peanut genotypes under end-of –Season drought conditions. Acta Physiol. Plant 29, 563–566.

Leloir, L.F., Goldenberg, S.H., 1960. Synthesis of glycogen from uridine diphosphate glucose in liver. Biol. Chem. 235, 919–923.

Liu, C.H., Liu, W., Lu, X.Z., Chen, W., yang, J.B., Zheng, L., 2014. Non destructive determination of transgenic bacillus thuringensis rice seeds (*Oryza sativa* L) suing multispectral imaging and chemometric methods. Food Chem. 153, 87–93.

Lorimer, G.H., Badger, M., Andrews, T.J., 1976. The activation state of ribulose-1,5-bisphosphate carboxylase by CO_2 and magnesium and physiological implications. Biochemistry 15, 529–536.
Lowry, O.H., Rosebrough, N., Farr, A.L., Randall, R., McCready, R.M., Guggloz, J., Silviera, V., Owens, H.S., 1950. Determination of starch and amylose in vegetables. Anal. Chem. 22, 1156–1158.
Lowry, O.H., Rosebrough, N.J., Farr, A.L., Randall, R.J., 1951. Protein measurement with folin-phenol reagent. J. Biol. Chem. 193, 265–275.
Luck, H., 1974. In: Bergmeyer (Ed.), Methods in Enzymatic Analysis 2. Academic Press, New York, p. 885.
Madamanchi, N.R., Donahue, J.V., Cramer, C.I., Alscher, R.G., Pedersen, K., 1994. Differential response of CuZn superoxide dismutase in two pea cultivars during a short-term exposure to sulphur dioxide. Plant Mol. Biol. 26, 95–103.
Malik, C.P., Singh, M.B., 1980. Plant Enzymology and Histoenzymology. Kalyani Publishers, New Delhi, 53.
Max, J.F.J., Schurr, U., Tantau, H.J., Hofmann, T, Ulbrich, A., 2012. Green house cover technology. Hortic. Rev. 40, in press.
Merah, O., Deléens, E., Monneveux, P., 1999. Grain yield and carbon isotope discrimination, mineral and silicon content in durum wheat under different precipitation regimes. Physiol. Plantarum 107, 387–394.
Miller, G.L., 1972. Use of DNS reagent for the determination of glucose. Anal. Chem. 31, 426.
Min, T.G., Kang, W.S., 2011. Simple, quick and non-destructive method for brassicaceae seed viability measurement with single seed base using resazurin. Horticulture, Environment and biotechnology 52(3), 240–245.
Misra, P.S., Mertz, E.T., Glover, D.V., 1975. Studies on corn proteins. VIII. Free amino acid content of Opaque-2 double mutants. Cereal Chem. 52, 844–848.
Moore, S., Stein, W.H., 1948. Photometric ninhydrin method for use in the chromatography of amino acids. J. Biol. Chem. 176, 367–388.
Morgan, J.M., 2000. Increases in grain yield of wheat by breeding for an osmoregulation gene: relationship to water supply and evaporative demand. Austral. J. Agric. Res. 51, 971–978.
Mukherjee, S.P., Choudhari, M.A., 1983. Implication of water stress-induced changes in the levels of endogenous ascorbic acid and hydrogen peroxide in *Vigna* seedlings. Physiol. Plant 58, 166–170.
Munns, R., James, R.A., 2003. Screening methods for salinity tolerance: a case study with tetraploid wheat. Plant Soil 253, 201–218.
Neumann, D., Nover, L., Parthier, B., Reiger, R., Scharf, K.D., Wollegeihn, R., Zur Neiden, U., 1989. Heat shock and other stress response systems in plants. Biol. Zentralba 108, 1–156.
O'Leary, M.H., 1981. Carbon isotope fractionation in plants. Phytochemistry 20, 53–567.
Passioura, J.B., 1986. Resistance to drought and salinity: avenues for improvement. Austral. J. Plant Physiol. 13, 191–201.
Passioura, J.B., 1996. Drought and drought tolerance. Plant Growth Regul. 20, 79–83.
Pateman, J.A., 1969. Regulation of synthesis of glutamate) dehydrogenease and glutamine synthetase in microorganisms. Biochem. J. 115, 769.
Pathak, P.S., 2000. Agro forestry: a tool for arresting land degradation. Indian Farming 49 (11), 15–19.
Pearcy, R.W., Ehleringer, J., Mooney, H.A., Rundel, 1989. Field methods and instrumentation. Plant Physiol. Ecol.

Peet, M.M., Willits, D.H., 1998. The effect of night temperature on greenhouse grown tomato yields in warm climate. Agric. Forest Meteorol. 92, 191–202.

Poorter, H., Climent, J., Van Dusschoten, D., Buhler, J., Postma, J., 2015. Pot size matters: a meta-analysis on the effects of rooting volume on plant growth. Funct. Plant Biol. 39, 839–850. doi:10.1071/FP12049

Popper, K.R., 1959. The logic of scientific discovery. Hutchinson, London.

Pranusha, P., Raja Rajeswari, V., Sudhakar, P., Latha, P., Mohan Reddy, D., 2012. Evaluation of groundnut genotypes for intrinsic thermo tolerance under imposed temperature stress conditions. Legume Res. 35 (4), 345–349.

Price, A.H., Hendry, G.A.F., 1991. Iron catalyzed oxygen radical formation and its possible contribution to drought damage in nine native grasses and three cereals. Plant Cell Environ. 14, 477–484.

Putter, J., 1974. In: Bergmeyer (Ed.), Methods of Enzymatic Analysis 2. Academic Press, New York, p. 685.

Ranganna, S., 1976. Manual of Analysis of Fruits and Vegetable Products. McGraw-Hill, New Delhi, 77–96.

Rao, R.C.N., Wright, G.C., 1994. Stability of the relationship between specific leaf area and carbon isotope discrimination across environments in peanut. Crop Sci. 34, 98–103.

Rao, R.C.N., Talwar, H.S., Wright, G.C., 2001. Rapid assessment of specific leaf area and leaf nitrogen content in peanut (*Arachis hypogaea* L.) using a chlorophyll meter. J. Agronomy Crop Sci. 189, 175–182.

Rao, R.C.N., Udayakumar, M., Farquhar, G.D., Talwar, H.S., Prasad, T.G., 1995. Variation in carbon isotope discrimination and its relationship to specific leaf area and ribulose-1,5-bisphosphate carboxylase content in groundnut genotypes. Austral. J. Plant Physiol. 22, 545–551.

Renuka Devi, K., Sudhakar, P., Sivashankar, A., 2013. Screening of paddy genotypes for high WUE and yield components. Bioinfolet 10 (113), 214–224.

Repo, T., Paine, D.H., Taylor, A.G., 2002. Electrical impedance spectroscopy in relation to seed viability and moisture content in snap bean. *Phaseolous vulgaris* L 12, 17–29.

Reporter, M., 1987. Nitrogen Fixation. In: Photosynthesis, J., Coombs, D.O., Hall, S.P., Long, J.M.O., Scurlock (Eds.), Techniques in Bioproductivity. Pergamon Press, Oxford, pp. 162–164.

Reynolds, M.P., Pask, A.J.D., Mullan, D.M. (Eds.), 2012. Physiological Breeding I: Interdisciplinary Approaches to Improve Crop Adaptation. CIMMYT, Mexico, D.F.

Romheld, V., 1991. The role of phytosiderophores in acquisition of iron and other micronutrients in graminaceous species: an ecological approach. Plant Soil 130, 127–134.

Ronchi, A., Garina, C., Gozzo, F., Tonelli, C., 1997. Effects of thiozolic fungicide on maize plant metabolism: Modification of transcript abundance in resistance-related pathways. Plant Sci. 130, 51–62.

Sadasivam, S., Manickam, A., 1992. Biochemical Methods for Agricultural Sciences. Wiley Eastern Ltd, New Delhi, India, 187–188.

Sadras, V.O., Reynolds, M.P., de la Vega, A.J., Petrie, P.R., Robinson, R., 2009. Phenotypic plasticity of yield and phenology in wheat, sunflower and grapevine. Field Crops Res. 110, 242–250.

Santamaria, J.M., Ludlow, M.M., Fukai, S., 1990. Contribution of osmotic adjustment to grain yield in Sorghum bicolor (L.) under water-limited conditions. I. Water stress before anthesis. Aust. J. Osmotic Adjustment Rice. Crop Sci. 20:310–314. Agric. Res. 41: 51–65.

Scharf, K.D., Hohfeld, I., Nover, L., 1998. Heat shock response and heat stress transcription factors. J. Biosci. 23, 313–329.

Scholander, P.E., Hammel, H.T., Hommingsen, F.A., Bradstreet, E.D., 1965. Sap pressure in vascular plants: negative pressure can be measured in plants. Science 148, 339–346.

Schopfer, P., Plachy, C., Frahry, G., 2001. Release of reactive oxygen intermediates (superoxide radicals, hydrogen peroxide and hydroxyl radicals) and peroxidase in germinating seeds controlled by light, gibberellins and abscisic acid. Plant Physiol. 125, 1591–1602.

Senthil-Kumar, M., Srikanthbabu, V., Mohan Raju, B., Ganeshkumar, Sivaprakash, N., Udayakumar, M., 2003. Screening of inbred lines to develop a thermotolerant sunflower hybrid using the temperature induction response (TIR) technique: a novel approach by exploiting residual variability. J. Exp. Bot. 54 (392), 2569–2578.

Servaites, J.C., Torisky, R.S., Chao, S.F., 1984. Diurnal changes in ribulose-1,5-bisphosphate carboxylase activity and activation state in leaves of field grown soybeans. Plant Sci. Lett. 35, 115–121.

Sheshshayee, M.S., Bindumadhava, H., Rachaputi, N.R., Prasad, T.G., Udayakumar, M., Wright, G.C., Nigam, S.N., 2006. Leaf chlorophyll concentration relates to transpiration efficiency in peanut. Ann. Appl. Biol. 148, 7–15.

Sheshshayee, M.S., Bindumadhava, H., Ramesh, R., Prasad, T.G., Lakshminarayana, M., Udayakumar, M., 2005. Oxygen isotope enrichment ($\Delta^{18}O$) as a measure of time-averaged transpiration rate. J. Exp. Bot. 56 (422), 3033–3039.

Sheshshsayee, M.S., Bindumadhava, H., Ramesh, R., Prasad, T.G., Udayakumar, M., 2010. Relationship between oxygen isotope enrichment ($\Delta^{18}O$) in leaf water, biomass stomatal conductance. Isotopes Environ. Health Stud. 46, 122–129.

Shindy, W.W., Smith, O.E., 1975. Identification of plant hormones from cotton ovules. Plant Physiol. 55, 550–554.

Silva Fernandes, A.M., Baker, E.A., Martin, J.T., 1964. Studies on plant cuticles. Ann. Appl. Biol. 53, 43–58.

Smillie, R.M., Hetherington, S.E., 1990. Screening for salt tolerance by chlorophyll flouroscence. In: hashimoto, Y. et al., (Ed.), Measurement Techniques in Plant Sciences, ninth ed. Academic Press, San Deigo, pp. 229–261.

Somogyi, M., 1952. Notes on sugar determination. J. Biol. Chem. 195, 19–23.

Specht, J.E., Williams, J.H., Pearson, D.R., 1985. Near-isogenic analyses of soybean pubescence genes. Crop Sci. 25(1).

Sreekanth, A., 1998. Thermotolerant groundnut (Arachis hypogaea L) genotypes identified based on temperature induction response (TIR) technique also exhibited enhanced expression of a few stress responsive proteins. MSc (Agriculture) thesis submitted to the University of Agricultural Sciences, Bangalore, India.

Sternberg, L.S.L., De Niro, M.J., Savidge, R.A., 1986. Oxygen isotope exchange between metabolites and water during biochemical reactions leading to cellulose synthesis. Plant Physiol. 82, 423–427.

Stobiecki, M., Skirycz, A., Kerhoas, L., Kachlicki, P., Muth, D., Einhorn, J., Mueller-Roeber, B., 2006. Profiling of phenolic glycosidic conjugates in leaves of Arabidopsis thaliana using LC/MS. Metabolomics 2 (4), 197.

Sudhakar, P., Babitha, M., Latha, P., Prasanthi, L., Reddy, P.V., 2006. Thermostability of cell membrane and photosynthesis in blackgram genotypes differing in heat tolerance. J. Arid Legumes 3 (2), 11–16.

Sudhakar, P., Latha, P., Babita, M., Prasanthi, L., Reddy, P.V., 2006. Physiological traits contributing to grain yields under drought in black gram and green gram. Indian J. Plant Physiol. PP 11, 391–396.

Sudhakar, P., Latha, P., Muneendrababu, A., 2010. Evaluation of sugarcane genotypes for high water use efficiency and thermo stability tolerance under imposed moisture stress at formative stage. Sugar Tech. 12 (1), 72–75.

Sudhakar, P., Latha, P., Ramesh Babu, P., Sujatha, K., Raja Reddy, K., 2012. Identification of thermotolerant rice genotypes at seedling stage using TIR technique in pursuit of global warming. Indian J. Plant Physiol. 17 (2), 185–188.

Sun, W., Bernard, C., van de Cotte, B.M., Van Montagu, M., Verbruggen, N., 2001. At-HSP17.6A, encoding a small heat shock protein in *Arabidopsis* can enhance osmotolerance upon overexpression. Plant J. 27, 407–415.

Swaminathan, M.S., 2005. Towards an ever-green revolution. In: Tuberosa, R., Phillips, R.L., Gale, M. (Eds.), In: Proceedings of the International Congress: In the Wake of the Double Helix: From the Green Revolution to the Gene Revolution 27–31. Avenue Media, Bologna, Italy, pp. 25–36.

Takebe, M., Yoneyama, T., Inada, K., Murakam, T., 1990. Spectral reflectance of rice canopy for estimating crop nitrogen status. Plant Soil 122, 295–297.

Tandon, H.L.S. Ed., 1993. Methods of analysis of soils, Plants, waters and fertilizers. Fertilisers Development and Consultation Organisation, New Delhi, India. 144.

Tang, G.Q., Luscher, M., Sturm, A., 1999. Antisense repression of vacuolar and cell wall invertase in transgenic carrot alters early plant development and sucrose partitioning. Plant Cell 11, 177–189.

Tangpremsri, T., Fukai, S., Fische, K.S., 1995. Growth and yield of sorghum lines extracted from a population for differences in Philippines. Osmotic adjustment. Aust. J. Agric. Res. 46, 61–74.

Tempest, D.W., Meers, J.C., Brown, C.M., 1970. Synthesis of glutamate in aerobactor aerogenes by a hitherto known roots. Biochem. J. 117, 405.

Thomas, H., Smart, C.M., 1993. Crops that stay green. Ann. Appl. Biol. 123, 193–219.

Towill, L.E., Mazur, P., 1975. Studies on the reduction of 2,3,5-triphenyltetrazolium chloride as a viability assay for plant tissue culture. Canad. J. Botany 53, 1097–1102.

Tuberosa, R., Salvi, S., 2004. QTLs and genes for tolerance to abiotic stress in cereals. In: Gupta, P.K., Varshney, R. (Eds.), Cereal Genomics. Kluwer, Dordrecht, The Netherlands, pp. 253–315.

Turner, N.C., Wright, G.C., Siddque, K.H.M., 2001. Adaptation of grain legumes (pulses) to water limited environments. Adv. Agronomy 71, 193–231.

Udayakumar, M., Bhojaraja, R., Sheshshayee, M.S., Gopalakrishsna, R., Jacob, J., 1999. How do plants cope with excess light? The role of ferritin. J. Plant Biol. 26, 135–142.

Udayakumar, M., Rao, R.C.N., Wright, G.C., Ramaswamy, G.C., Ashok Roy, S., Gangadhar, G.C., Aftab Hussain, I.S., 1998. Measurement of transpiration efficiency in field condition. J. Plant Biol. 1, 69–75.

Vasanthi, R.P., Reddy, P.V., Jaya lakshmi, V., Sudhakar, P., Asalatha, M., Sudhakar Reddy, P., Harinatha Naidu, P., Muralikrishna, T., Venkateswarulu, O., John, K., Basu, M.S., Nigam, S.N., Nageswara Rao, R.C., Wright, G.C., 2006. A high-yielding drought-tolerant ground variety Abhaya. Int. Arachis Lett. 26, 15–16, 2006.

Vega, J.M., Jacobo, C., Manuel, L., 1980. In: Anthony San Pietro (Ed.), Methods in Enzymology. Academic Press, New York, 255.

Venuprasad, R., Shashidhar, H.E., Hittalmani, S., Hemamalini, G.S., 2002. Tagging quantitative trait loci associated with grain yield and root morphological traits in rice (*Oryza sativa* L) under contrasting moisture regimes. Euphytica 128, 293–300.

Verma, D.P.S., 1999. Osmotic stress tolerance in plants: role of proline and sulphur metabolism. In: Shinozaki, K., Yamaguchi Shinozaki, K. (Eds.), Molecular Response to Cold Drought Heat Salt Stress in Higher Plants. R. G. lands Company, Texas, USA, pp. 153–168.

Watson, D.J., 1952. The physiological basis of variation in yield. Advances in Agronomy.

Williams, R.F., 1964. The quantitative description of growth. In: Barnard, C (Ed.), Grasses and Grasslands. Macmillan, London, pp. 89–101.

Wogan, G.N., 1999. Aflatoxin as a human carcinogen. Hepatology 30, 573–575.

Wright, G.C., Hammer, L., 1994. Distribution of nitrogen and radiation use efficiency in peanut canopies. Austral. J. Agric. Res. 45, 565–574.

Wright, G.C., Rao, R.C.N., Farquhar, G.D., 1994. Water-use efficiency and carbon isotope discrimination in peanut under water deficit conditions. Crop Sci. 34, 92–97.

Xia Xin, Wan, Y., Wang, W., Yin, G., McLamore, E.S., Lu, X., 2013. A real-time, non-invasive, microoptrode technique for detecting seed viability by using oxygen influx. Scientific Reports 3, Article number: 3057: Nature publishing group.

Zeevaart, J.A.D., 1980. Change in the level of abscisic acid and its metabolites in excised leaf blades of *Xanthium strumarium* during and after stress. Plant Physiol. 66, 672–678.

Index

Q

R

S